Essential Smart Growth Fixes for Urban and Suburban Zoning Codes

ACKNOWLEDGEMENTS

Principal author:
Kevin Nelson, AICP, US EPA

Contributing authors:
Amy Doll, ICF International
Will Schroeer, ICF International
Jim Charlier, Charlier Associates
Victor Dover, Dover, Kohl & Partners
Margaret Flippen, Dover, Kohl & Partners
Chris Duerksen, Clarion Associates
Lee Einsweiler, Code Studio
Doug Farr, Farr Associates
Leslie Oberholtzer, Farr Associates
Rick Williams, Van Meter Williams Pollack

Contributors and reviewers from US EPA:
John Frece
Abby Hall
Lynn Richards
Megan Susman

Document Layout:
Colin Scarff, Code Studio

Additional Participants: January 2008 and October 2008 Workshops:

US EPA:
Geoff Anderson
Kevin Nelson
Ilana Preuss
Lynn Richards
Tim Torma

Center for Planning Excellence (host of January 2008 workshop)
Elizabeth "Boo" Thomas
Camille Manning-Broome

CONTENTS

INTRODUCTION

Across the country, state and local governments are searching for ways to create vibrant communities that attract jobs, foster economic development, and are attractive places for people to live, work, and play. Increasingly, these governments are seeking more cost-effective strategies to install or maintain infrastructure, protect natural resources and the environment, and reduce greenhouse gas emissions. What many are discovering is that their own land development codes and ordinances are often getting in the way of achieving these goals.

Fortunately, there is interest in tackling these challenges. As the nation's demographics change, markets shift, and interest in climate change, energy efficiency, public health, and natural resource protection expands, Americans have a real opportunity to create more environmentally sustainable communities.

To address these issues, many local governments want to modify or replace their codes and ordinances so that future development and redevelopment will focus on creating complete neighborhoods—places where residents can walk to jobs and services, where choices exist for housing and transportation, where open space is preserved, and where climate change mitigation goals can be realized. Many local governments, however, lack the resources or expertise to make the specific regulatory changes that will create more sustainable communities. And for many, model codes or ordinances can be too general for practical use or are often designed to be adopted wholesale, which many communities are unprepared to do.

To respond to this need, the U.S. Environmental Protection Agency's (EPA) Development, Community, and Environment Division (DCED), also known as the Smart Growth Program, has put together this document to help those communities that may not wish to revise or replace their entire system of codes and ordinances, but nevertheless are looking for "essential fixes" that will help them get the smarter, more environmentally responsible, and sustainable communities they want.

Smart growth creates lively walkable places that bring businesses to the street.

Farr Associates

To find the changes that can be most helpful, DCED convened a panel[1] of national smart growth code experts to identify what topics in local zoning codes are essential to creating the building blocks of smart growth. This document presents the initial work of that panel. It is an evolving document, one that will be regularly revised, added to, and updated. It is intended to spark a larger conversation about the tools and information local governments need to revise their land development regulations.

The purpose of this document is to identify the most common code and ordinance barriers communities face and to suggest actions communities could take to improve their land development regulations. Given the effort and political will that is necessary to make any changes to local regulations, the suggested code provisions are separated into three categories:

- **Modest Adjustments:** Code suggestions in this category assume the local government will keep the existing regulations and is looking for relatively modest revisions that will help it remove barriers to building smart growth developments or create a regulatory framework where all development types are on equal footing. Examples include changing code language from minimum setbacks or parking requirements to maximums.

- **Major Modifications:** Code suggestions in this category assume the local government is looking to change the structure of the existing code. Suggestions include creating incentives for smart growth development or creating overlay zones and mixed-use districts.
- **Wholesale Changes:** Code suggestions in this category assume the local government wants to create a new regulatory framework, such as creating a form-based code or requiring sidewalks and alleys.

Every community is distinct, with different landscapes, natural resources, demographics, history, and political culture. Some communities have found that an incremental approach to code changes works best, while others have found success in wholesale change. This document strives to provide a starting point for all communities by recognizing their wide variability.

The document includes eleven Essential Fixes to the most common barriers local governments face when they want to implement smart growth approaches. Each Essential Fix describes the problem and how to respond, expected benefits, and implementation steps. Other resources include practice pointers and examples.

This tool does not include model language, nor is it intended to provide model codes or ordinances. The information here, however, can help communities evaluate their existing codes and ordinances and apply the information to achieve smart growth objectives. This document focuses primarily on barriers in suburban and urban communities. Similar issues regarding rural development will be addressed in a subsequent document that is under development. The intent is to continually revise, update, and expand the information provided here. Please send comments, feedback, or suggestions to the EPA project manager, Kevin Nelson, AICP, at nelson.kevin@epa.gov or 202-566-2835.

---

[1]   The panel met in January and October 2008. See the Acknowledgements for a list of participants.

ALLOW OR REQUIRE MIXED-USE ZONES

INTRODUCTION

A common problem with the conventional Euclidean zoning used by many communities is its focus on separating potentially incompatible land uses. This separation has made our development patterns inefficient, forcing residents to drive longer distances to get to their jobs, schools, shops, and services, which increases traffic congestion, air pollution, and greenhouse gas emissions. The underlying health and safety problems that zoning was designed to address 80 years ago—separating homes from factories, stock yards, and other "nox-ious" uses—are still important, but in our current economy, many commercial uses and workplaces can be integrated with homes without "noxious" effects. The health and safety goals of separating uses must now be placed in context with a range of other problems that are created by not allowing uses where they will be most efficient. Such separation can frustrate efforts to promote alternative modes of transportation and create lively urban places.

Rockville Town Square in Maryland contains a vibrant mixture of offices, residences, retail and gathering space for people to enjoy.

US Environmental Protection Agency

♻EPA United States Environmental Protection Agency

RESPONSE TO THE PROBLEM

The response to this problem is to encourage or require more mixed-use zones. Mixed-use zones will look different in various contexts, from downtowns to transit-oriented development (TOD) to commercial corridors to the neighborhood corner store. Communities should be mindful of these variations so that there is not a "one size fits all" solution for how land uses are mixed to accommodate market conditions and design expectations. Requiring vertically mixed-use buildings, such as a building with ground-floor retail and offices or residences in the upper floors, along older, pedestrian-oriented corridors can reinvigorate a sleepy street. Alternatively, simply permitting a variety of uses within one zoning district allows a horizontal mix of uses that can break up the monotony of single uses, such as strip centers or single-family housing. This horizontal mix can make a street more interesting and bring stores, services, and workplaces closer to residents.

Farr Associates

EXPECTED BENEFITS

- Reduction in vehicle miles traveled, resulting in lower greenhouse gas emissions, lower commuting costs, and decreased road congestion.

- More balanced transportation systems that support walking, bicycling, and public transit, as well as driving.

- Livelier urban spaces with public gathering places and a variety of shops, restaurants, and entertainment.

- Complete neighborhoods where residents can live, work, and play.

- Diversity of housing for people of all incomes and at all stages of life.

- More vibrant commercial areas that provide retail and services for patrons.

- More compact development that helps preserve open space in outlying areas by reducing the need and demand for low-density, sprawling development.

- Efficient use of services and infrastructure, resulting in cost savings for the public.

STEPS TO IMPLEMENTATION

1. Modest Adjustments

- Define mixed-use areas/activity centers in land use plans (on a neighborhood, community, and/or regional scale), and designate preferred locations for them.

- Permit residences in the upper floors of buildings in appropriate existing commercially zoned districts.

2. Major Modifications

- Remove obstacles to mixed-use development by creating zoning districts that allow mixed-use development by right (i.e., without the need for a rezoning or special discretionary approval process).

- Develop a variety of mixed-use districts, including vertical mixed uses and horizontal mixed uses, as needed. The context of uses (e.g., main street, neighborhood setting) is important for determining the type of mixed-use district.

- Designate mixed-use districts on the official zoning map.

3. Wholesale Changes

- Synchronize zoning codes and area plans to coordinate the location and development of mixed-use districts.

PRACTICE POINTERS

- Consider mandatory mixed-use development in preferred locations (e.g., near transit stops) to ensure that these prime locations are not used for low-density, single-use development.

- Adopt compatibility standards to ensure adequate transitions to adjacent, lower-density uses. Consider architectural, design, open space, operational, and other categories of transitional standards.

- Tailor development standards (such as parking, open space, and landscaping regulations) for mixed-use developments so as not to create unintended hurdles for this preferred development form. For example, typical parking requirements often do not reflect the reduced need for parking typical of most mixed-use developments. The additional land that such excessive standards require for parking can spread out growth so that lively, compact developments are hard to achieve.

- Use market studies to ensure an appropriate amount of commercially and residentially zoned land. Avoid requiring more vertically mixed uses than the market can support. Horizontal mixed-use districts can allow the market to determine the appropriate mix of uses. Establish standards for the development of each use within the area to ensure contiguous retail areas. In these locations, establish triggers such as achieving market benchmarks for renewed planning efforts as the area begins to change.

- Level the playing field for mixed-use developments. For example, make sure that single-use commercial strip developments are held to the same high design and other standards required of mixed-use developments.

- Create incentives for mixed-use development, such as a wider array of permitted uses in mixed-use districts (as opposed to single-use districts), increased densities, and accelerated application processing.

EXAMPLES AND REFERENCES

- International City/County Management Association and Smart Growth Network. *Getting to Smart Growth: 100 Policies for Implementation.* 2002. EPA 231-R-05-001. http://www.epa.gov/smartgrowth/getting_to_sg2.htm.

- Ewing, R., Bartholomew, K., Walters, J., Chen, D. *Growing Cooler: The Evidence on Urban Development and Climate Change.* Urban Land Institute. 2008. p. 25.

- Lewis, L. "Celebration Traffic Study Reaffirms Benefits of Mixed-Use Development." *Transportline.* HDR. 2004. http://www.hdrinc.com/Assets/documents/Publications/Transportline/September2004/CelebrationTrafficStudy.pdf.

- Coupland, A. *Reclaiming the City: Mixed Use Development. Routledge.* November 1996. p. 35.

- Williams, K. and Seggerman, K. *Model Regulations and Plan Amendments For Multimodal Transportation Districts.* Florida Department of Transportation. April 2004. pp. 7-14. http://www.dot.state.fl.us/planning/systems/sm/los/pdfs/MMTDregs.pdf.

- Oregon Transportation and Growth Management Program. *Commercial and Mixed-Use Development Code Handbook.* October 2001. pp. 33-38. http://egov.oregon.gov/LCD/docs/publications/commmixedusecode.pdf.

- Morris, M., ed. "Sec. 4.1: Model Mixed-Use Zoning District Ordinance." *Model Smart Land Development Regulations.* Interim PAS Report. American Planning Association. March 2006. pp. 3-5. http://www.planning.org/research/smartgrowth/pdf/section41.pdf.

- Duany Plater-Zyberk & Company. *SmartCode, Version 9.2.* February 2009. http://www.smartcodecentral.com/smartfilesv9_2.html.

- City of Colorado Springs, Colorado. *Mixed Use Development Design Manual.* March 2004. pp. 56-64. http://permits.springsgov.com/units/planning/Currentproj/CompPlan/MixedUseDev/IV-%20E.pdf.

USE URBAN DIMENSIONS IN URBAN PLACES

INTRODUCTION

Conventional zoning codes are typically replete with various dimensional standards that govern a range of topics, including minimum lot sizes and widths, floor area ratios, setbacks, and building heights. These standards are generally geared to produce low-intensity, low-rise residential and commercial development. Even codes for more mature urban areas often reflect this lower-density orientation. While this development pattern may be appropriate in some areas and under some circumstances (e.g., around environmentally sensitive areas), these standards often have unintentionally stifled more compact development in many cities and towns, preventing the development of attractive, lively, and cost-efficient places. Recalibrating dimensional standards can help accommodate and promote a more compact development pattern and create attractive urban environments. Changes in dimensional standards can also improve connectivity enhanced site planning and design. (See Essential Fixes Nos. 4 and 6 for street- and parking-related dimensional standards.)

This street in the Georgetown neighborhood of Washington, DC exhibits a mature development of a city street.

RESPONSE TO THE PROBLEM

Cities across the country have been built based on the availability of land and proximity to jobs and amenities. Dimensional standards were established to accommodate these conditions. As communities and prosperity yielded larger lots and more spread-out development, communities began to reassess their function and design. A compact, walkable neighborhood is achieved through design and direction from codes and ordinances. A principal way of creating this type of place is through modifications to the dimensional standards—that is, the size of lots, setback requirements, height restrictions, and the like.

Form-based codes are a typical response for communities that are looking to increase options for compact form and walkable neighborhoods. Components of form-based codes include regulating plans, building form standards (building siting and height), and optional architectural elements. In essence, the form of the building is more important than the use that occupies it.

EXPECTED BENEFITS

- More compact development patterns that help preserve open space in outlying areas.

- Higher density development that supports transit and mixed-use activity centers.

- A more attractive public realm that is designed to balance pedestrians and bicyclists with the car.

- Cost-efficient provision of infrastructure and services.

STEPS TO IMPLEMENTATION

1. Modest Adjustments

- Tailor dimensional standards in the development code to promote more compact development. Consider changing minimum standards to maximums.

  – For residential development, relevant changes could include lot width and area changes, smaller yards, increased lot or building coverage for smaller lots, increased height, and increased density.

  – For commercial or mixed-use development, relevant changes could include increased height, smaller yards and open space, increased lot or building coverage, and increased floor area ratios (FAR).

- Replace FAR with form standards such as height and maximum setbacks. Consider limiting building footprints in neighborhood commercial areas.

- Modify codes for commercial districts to allow residential development, especially over first-floor retail.

- Eliminate landscape buffers in the commercial area; there is no need to buffer like uses, such as two office buildings or a restaurant and a store, from each other.

2. Major Modifications

- Create incentives to provide multiple housing types in existing districts through dimensional standards (e.g., enable small lots and limited buffer yards between homes).

- Establish or reduce block lengths or perimeters to produce better connections and increase walkability.

US Environmental Protection Agency

Pedestrians traverse through a neighborhood park to reach homes and businesses that are built to the street line, creating appropriate dimensions for common open space amidst small lots.

- Adopt context-based or neighborhood-based dimensional standards that replicate existing, appealing, compact neighborhood patterns (e.g., narrow street width, sidewalks wide enough for safe and comfortable walking).

- Revise the codes for existing districts to encourage neighborhood redevelopment by applying new dimensional standards such as smaller lot requirements.

- Create districts for new compact building and development types that are not currently found in your community or neighborhood. (See the discussion of mixed use in Essential Fix No. 1.)

3. Wholesale Changes

- Coordinate new form-based dimensional standards, such as the siting of buildings, with zoning map changes to reflect the nature of form-based development versus use-specific zones.

- Plan a subarea of the community, then develop or calibrate and adopt a form-based code to create an option for additional compact, walkable neighborhoods.

PRACTICE POINTERS

- Where significant change in dimensional standards is proposed, create a computer model, preferably in 3-D (using ArcGIS or a similar program), of the existing standards in comparison to the proposed standards.

- Consider design and operational compatibility standards to ensure that new compact development is compatible with surrounding lower-density residential neighborhoods.

- Revise subdivision specifications and standards (e.g., narrower streets, reduced minimum driveway width) to encourage denser, more compact development.

- Relate dimensional standards to the transportation system (e.g., modify setbacks based on right of way instead of the street width).

- Replace standards that allow a variety of forms, such as FAR, with ones that provide a consistent benchmark, such as height requirements.

- Include other agencies, such as the public works or fire departments, early in discussions regarding efforts to revise dimensional standards.

- Analyze stormwater management requirements of denser developments, and consider green infrastructure approaches. (See Essential Fix No. 9.)

EXAMPLES AND REFERENCES

- Oregon Transportation and Growth Management Program. *Commercial and Mixed-Use Development Code Handbook.* October 2001. pp. 40-43. http://egov.oregon.gov/LCD/docs/publications/commmixedusecode.pdf.

- Freidman, S.B. and American Planning Association. *Planning and Urban Design Standards.* John Wiley and Sons. April 2006. pp. 664-666.

- City of Franklin, Tennessee. "Chapter 5: Dimensional Standards." *City of Franklin Zoning Ordinance.* http://www.franklintn.gov/planning/Side-by-Side%20Comparison%20Workshops/Chapter%205/Side-by-side%20Comparision%20Ch%205-%20Part%20One.pdf. Accessed August 12, 2009.

- City of Durham, North Carolina. *Durham City-County Unified Development Ordinance.* http://www.durhamnc.gov/udo. Accessed August 12, 2009.

- City of Colorado Springs, Colorado. *Mixed Use Development Design Manual.* pp. 56-64. March 2004. http://permits.springsgov.com/units/planning/Currentproj/CompPlan/MixedUseDev/IV-%20E.pdf.

- U.S. Green Building Council. LEED for Neighborhood Development (LEED-ND). http://www.usgbc.org/leed/nd. Accessed May 15, 2009.

- Parolek, D. et al. *Form-Based Codes: A Guide for Planners, Urban Designers, Municipalities and Developers.* John Wiley and Sons, Inc.: New Jersey. 2008. pp. 12-17.

3 REIN IN AND REFORM THE USE OF PLANNED UNIT DEVELOPMENTS

INTRODUCTION

The inflexibility of Euclidean single-use zone districts, inappropriate development and dimensional standards, and Byzantine approval processes have given rise to the use of negotiated developments in many communities. These negotiated developments usually take the form of planned unit developments (PUDs), planned developments, or master-planned communities. This discussion will use PUD as the collective term. PUDs allowed communities to overcome some of the strictures of Euclidean zoning and provided a vehicle for local government to negotiate community benefits such as additional open space, recreational facilities, better design, and contributions to infrastructure. PUDs, which spread rapidly after the concept was introduced in the 1960s, are attractive because they are often simpler and quicker than seeking multiple amendments and variances to an outdated zoning code.

Originally, PUDs were conceived of and used to allow flexibility in design standards to take advantage of site characteristics or to address community goals (e.g., clustering development to provide open space or protect sensitive natural areas). PUDs were meant to achieve higher quality developments and meet community goals better than the standard subdivision and

New Town in St. Charles, Missouri features is a planned unit development that encapsulates a variety of smart growth and new urbanism features including compact development, mix of land uses and design guidelines to create a distinctive place.

zoning regulations would allow. Sea Ranch in Northern California was a model of PUD, using attractive design to better integrate with the natural environment. Many of the initial Traditional Neighborhood Developments (TNDs) were approved through a PUD process.

Today, however, relatively standard subdivisions are being approved using PUDs as an alternative to rewriting zoning and subdivision regulations for time and cost considerations. PUDs allow communities to impose conditions as part of the approval, which cities use to ensure they receive the appropriate infrastructure, off-site improvements, and fees to offset development impacts. The initial objective of distinctive or attractive design, however, often is lost as part of the PUD process.

The PUD approach has now proliferated to the point that most projects of any size or significance are approved that way. Some observers estimate that upwards of 40 percent[2] of all residential units in the United States each year are approved through a PUD process, not conventional zoning. The result is that many growing cities are not the products of their land use plans and zoning codes, but rather the result of individually negotiated agreements. Indeed, in a growing number of communities, all major developments are being reviewed through the PUD process.[3]

As this trend proliferates, communities have increasingly recognized the downside of relying too heavily on PUDs and negotiated developments, including:

- There is significant uncertainty for developers, who have no standards to guide the development approval process, and for neighbors of proposed PUDs, who find that they cannot rely on existing zoning or land use plans and that the city planning staff controls much of the planning process.
- Project reviews can become longer, less efficient, and politically charged and can drag out for years.

- Major planning decisions are made with less public input into defining the community objectives prior to a development proposal.
- Environmental and design standards are often minimized in the process.
- Often this process creates an administrative nightmare for staff that have to deal with multiple mini-zoning codes created for each PUD, each of which differs on development standards and other requirements.
- The planning process becomes a project-by-project process rather than a comprehensive development review, and more of a political process than an evaluation of planning regulations and community goals.

RESPONSE TO THE PROBLEM

To respond to these problems, communities are reducing the use of PUDs by updating their zoning districts and standards to accommodate preferred development patterns and types. They are also limiting the use of PUDs to larger projects that can provide compensating community benefits without waiving key design and environmental standards.

Communities are attempting to get out in front of PUD proposals by creating PUD zoning regulations or design guidelines. These are generally developed as part of a community design process so that the city can define its goals for a site or area prior to specific development proposals. Principles, regulations, and design guidelines are then used in conjunction with PUD zoning to provide clearer direction while allowing the desired design flexibility.

2   Duerksen, C. "Rural Smart Growth Zoning Code Tools." American Planning Association National Conference, April 28, 2009.

3   Ibid.

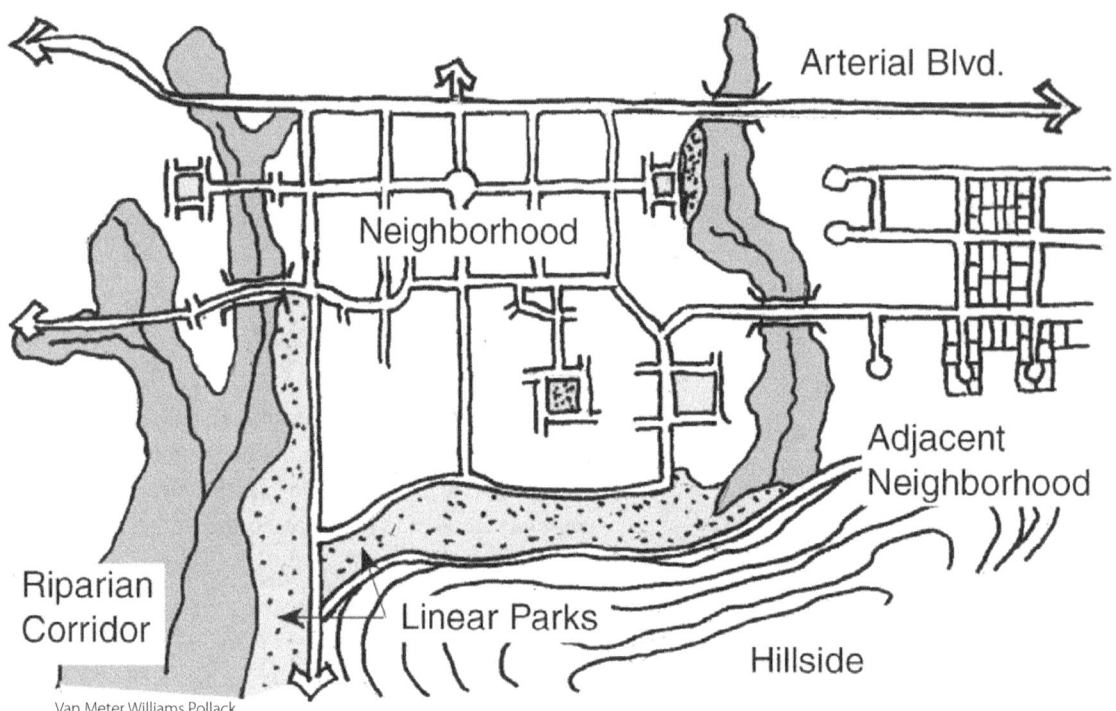

This drawing of the Belmar neighborhood shows how the development fits within the context of neighboring uses.

Van Meter Williams Pollack

EXPECTED BENEFITS

- Increased certainty and predictability in the development review process while still allowing appropriate design flexibility.

- Setting the basic goals and fundamental standards for an area's development prior to a specific development proposal:

  – Creates an efficient design and review process and requires less staff time to administer the development over time.

  – Adheres to community growth visions and goals as established in comprehensive plans and gives the development sector clear direction on the quality, character, and fundamental elements the community wishes to see in any proposal.

  – Prevents important design and environmental standards from being waived or weakened in the PUD process.

STEPS TO IMPLEMENTATION

1. Modest Adjustments

- Reform the PUD process to ensure that the parcel is designed appropriately given topography, adjacent uses, and additional impacts in the PUD-designated areas, and reduce the use of PUDs on small sites (under 2 acres).

- Remove or substantially reduce the need to use PUDs by fixing dimensional standards, particularly on small parcels. (See Essential Fix No. 2.)

- Create standards for PUD (e.g., apply Traditional Neighborhood Design policies, standards, and design guidelines as base PUD regulations prior to receiving development proposals).

- If PUDs are allowed, rein them in by establishing a minimum size for PUD projects, identifying specific allowable locations, and prohibiting waivers or other weakening of important environmental and design standards.

2. Major Modifications

- Prohibit PUDs as an alternative to following comprehensive plans and zoning codes. This may require communities to run public input processes to provide the detailed goals, objectives, and design elements for individual development proposals for larger sites. The community may also decide to rewrite its zoning regulations.

3. Wholesale Changes

- Create distinctive area and sector plans that give clear guidance to staff and the development community as to the vision and intended built-out of development. Complement these plans with accompanying zoning.

- Prior to accepting a development proposal for an area, communities should undergo a public master planning process to set goals and objectives; map land use and zoning; and set standards, regulations, and development quality through guidelines for the entire planning area.

- Implement an overlay district that allows the development of a site or area if specific standards are adopted. An example could be an overlay of the SmartCode or another set of development regulations onto an area designated in the comprehensive plan for future development.

PRACTICE POINTERS

- Consider establishing a list of compensating community benefits (such as a park, sidewalks, or trails) that the community expects in return for flexibility in uses, density, and other factors. This will reassure the community that they will get benefits from development and provide some certainty for developers regarding negotiated benefits.

EXAMPLES AND REFERENCES

- Newby, B. "Planned Unit Development: Planning Implementation Tools." Center for Land Use Education. November 2005. ftp://ftp.wi.gov/DOA/public/comprehensive-plans/ImplementationToolkit/Documents/PUD.pdf.

- New York State Legislative Commission on Rural Resources. *A Guide to Planned Unit Development.* State of New York. Fall 2005. pp. 4-8. http://www.dos.state.ny.us/lgss/pdfs/PUD1.pdf.

- Benton County, Oregon. "Chapter 100: Planned Unit Development in Corvallis Urban Fringe." *Benton County Development Code.* April 1999. http://www.co.benton.or.us/cd/planning/documents/dc-ch_100.pdf. Accessed August 12, 2009.

- City of Westminster, Colorado. *Design Guidelines for Traditional Mixed Use Neighborhood Developments.* April 2006. pp. 12-18. http://www.ci.westminster.co.us/files/tmund.pdf.

- City of Mountain View, California. "Precise Plans." http://www.mountainview.gov/city_hall/community_development/planning/plans_regulations_and_guidelines/precise_plans.asp. Accessed August 12, 2009.

- St. Lucie County, Florida. "Chapter 7: Recreation and Open Space Element." *Land Development Code.* May 2009. http://www.municode.com/resources/gateway.asp?pid=14641&sid=9. Accessed August 12, 2009.

- Larimer County, Colorado. "Proceedings of the Board of County Commissioners, February 8, 1999." http://www.co.larimer.co.us/bcc/1999/BC990208.HTM. Accessed July 10, 2009.

 # FIX PARKING REQUIREMENTS

INTRODUCTION

The parking standards found in many conventional zoning codes can be a significant barrier to lively, mixed-use developments and activity centers, especially in existing downtowns. Parking standards commonly in use in the United States often call for too much off-street parking and require all or too much of it to be provided on the development site. Also, many zoning codes do not allow consideration of alternative parking arrangements, such as shared parking or credit for on-street parking that can reduce the need for on-site spaces and help create a more attractive streetscape. Such regulations fail to recognize the difference between parking demand in various contexts.

Codes and regulations should enable adjacent uses to share parking as evidences by the demand or overlap in this chart.

Van Meter Williams Pollack

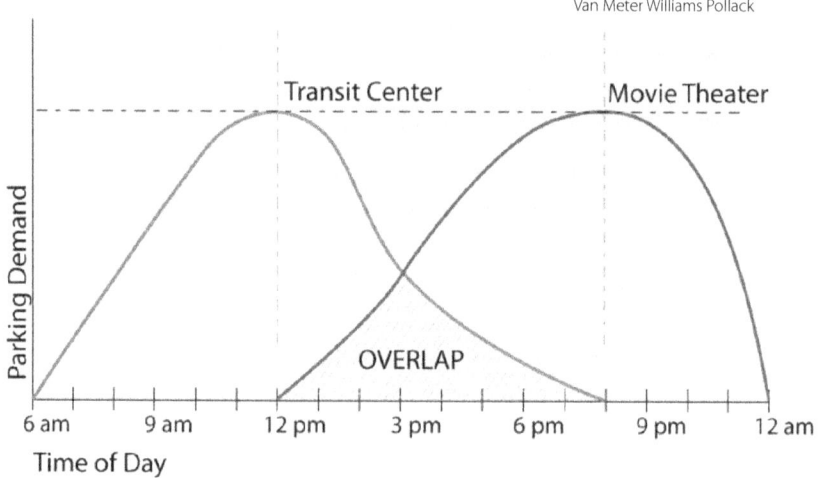

In many communities, the effect of conventional parking requirements is to make redevelopment of smaller parcels in older, mature areas infeasible and to make dense, compact, mixed-use development nearly impossible because of the code requirement for large expanses of surface parking or expensive structured parking. Large areas of surface parking in commercial areas discourage walking and actually increase parking demand by forcing people to drive between destinations. Frequently, zoning codes or development regulations allow (or even require) surface parking to be placed between buildings and the street, and they often allow parking structures to be built as stand-alone uses—both of which are deadly to vibrant, pedestrian-oriented places.

RESPONSE TO THE PROBLEM

Municipal governments across the country have been working to create more effective parking management systems for at least a couple of decades. The best parking management systems have these characteristics in common:

- They recognize that *too much* parking can be a serious issue, but so can *not enough* parking. Regulating parking supply became common in the first place because of the issues caused when developers provided inadequate parking and parking spilled over into nearby neighborhoods. What is generally needed is "the right amount" of parking, which can vary widely by place and by time. Good parking systems are carefully balanced to be specific to their settings and are adaptable to changes over time.

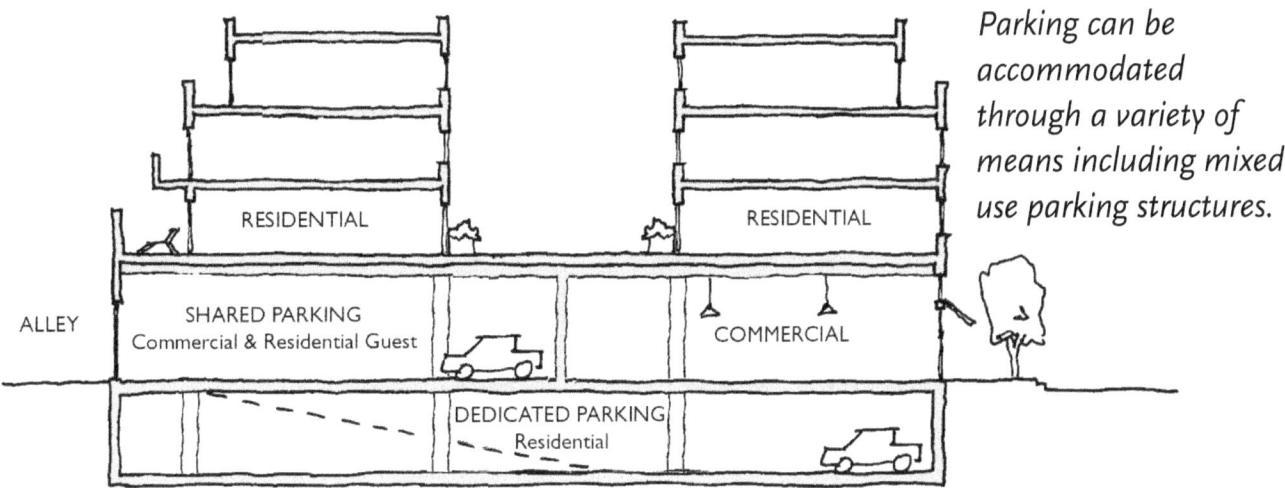

Parking can be accommodated through a variety of means including mixed use parking structures.

RESIDENTIAL | RESIDENTIAL

ALLEY | SHARED PARKING
Commercial & Residential Guest | COMMERCIAL

DEDICATED PARKING
Residential

Van Meter Williams Pollack

- They recognize that parking policy must be well integrated with overall transportation policy and land use policy. Transit services, good bicycle facilities, and a great walking environment can reduce parking demand significantly. Mixed-use development coupled with good walking environments can reduce parking demand even further. However, these transportation options must be in place before reducing parking requirements. For example, it makes little sense to reduce parking supply so that people will ride the bus if transit service levels are too low to attract ridership.

- They take into account that parking is inherently expensive. Surface parking consumes valuable land, removing it from productive use. Structured parking incurs capital costs that can exceed $20,000 per space,[4] thereby subtracting capital funds from development. Successful parking management systems reconcile the cost of providing parking with local taxation and fees, with the fine schedule for parking violations, and with the fees charged for use of parking.

Successful municipal parking management systems generally incorporate some combination of the following strategies and measures:

- **Lower Parking Supply Minimums** – The minimum parking requirements in many local codes are based on demand studies conducted in spread-out suburban places. These studies reflect parking demand in settings where shoppers and workers do not or cannot walk or use transit. In mixed-use settings with good pedestrian environments, such regulations overestimate parking demand and have a self-fulfilling effect by making mixed-use devel-

opment and redevelopment physically impossible.

- **Off-Site Parking** – In mixed-use environments, parking should be treated as a utility, not an on-site private activity. Requiring each landowner in a downtown to provide private parking on his or her parcel is akin to requiring each landowner to drill his or her own water well. Modern parking ordinances allow parking minimums to be met off site, although they may require that the parking location be within a maximum 600- to 1,000-foot distance from the development. These could be private joint parking facilities or public facilities owned by a parking district. The developer is still responsible for the cost of parking, either directly through capital fees or indirectly through property taxes. In some settings, it is feasible to "unbundle" parking from residential projects, allowing parking to be provided on the open market.

- **Fee-In-Lieu System** – In places where the city is providing public parking facilities or where a parking district has been created, provisions can be written that allow a developer to pay a set fee in lieu (FIL) of providing parking supply directly. The money from FIL payments is then used to expand public parking supply. It is important that any FIL fee schedule be realistic about actual costs of parking.

- **Shared Parking Credits** – Spread-out parking requirements assume that each business has its own separate parking supply and that it must be large enough to accommodate the peak hour of the peak day of the year. That assumption results in excessive parking. Different parking uses peak at different times of day—office parking in the middle of the day, retail in late afternoon and on week-

---

4   U.S. EPA. *Parking Spaces / Community Places: Finding the Balance Through Smart Growth Solutions.* February 2006. EPA 231-K-06-001. p. 9.

ends, restaurants in the evening. Shared parking provisions allow developers to reduce parking supply requirements when different uses can share the same parking spaces.

- **Parking Enforcement** – A pervasive cause of perceived parking shortages is the misuse of premium parking by employees. The closest, most convenient parking spaces—storefront, on-street parking in particular—should be protected for use by customers. Yet in many places, these spaces are occupied by employees' cars. Even where time restrictions have been established, they are often poorly enforced or the fines are too low to deter routine abuse. This situation can be corrected by ensuring there is adequate employee parking nearby and by adequately staffing enforcement.

- **Public Transit** – Many communities have reduced parking demand in mixed-use areas by improving transit service, especially for commuters. This approach is especially attractive because it reduces parking demand while improving mobility and access. Transit provides environmental benefits as well, including reduced air pollution and greenhouse gas emissions.

- **On-Street Parking** – The most valuable parking in most commercial and mixed-use places is parking on the street in front of businesses. Yet many cities are careless about keeping on-street parking or do not do enough to ensure the maximum number of spaces per block. Shifting from parallel to diagonal parking can increase parking supply by up to 30 percent per block face.

EXPECTED BENEFITS

- Lower cost of redevelopment and infill projects, helping them compete with outlying projects.

- Lively, active, economically strong mixed-use districts that are regional destinations.

- Increased tax base and tax revenues.

- Increased transit patronage that supports increased levels of transit service.

- More pedestrian-friendly environments.

STEPS TO IMPLEMENTATION

(Note: some of these measures are in support of code changes, but are not in themselves addressed through the zoning or land development code.)

1. Modest Adjustments

- Create a parking overlay district in the parking code for a downtown or other mixed-use area. Reduce minimum off-street parking supply requirements in the overlay district based on recalculated demand resulting from alternative transportation options, the mix of land uses, and a "park once" strategy that encourages parking in one place and walking to multiple destinations. Calculate a shared parking allowance based on the specific land uses in the overlay district.

- Develop residential parking permit provisions to help protect neighborhoods affected by overflow parking resulting from increased parking enforcement. Design the system to be applied in neighborhoods (not automatically citywide) based on criteria, such as the actual amount of on-street parking demand. Carefully manage and enforce the residential parking permit system to avoid abuse, such as sale of permits. Consider returning a portion of receipts from parking permit fees to the neighborhood in the form of street repairs and improvements. Consider selling "commuter permits" for residential streets in parking permit districts near mixed-use centers, with all or some of the revenue returned to the neighborhood in the form of capital repairs and improvements.

- Work with the public works department to increase the amount of on-street parking in a downtown or other mixed-use center. Convert parallel to diagonal parking where feasible. Evaluate parking stall specifications (length and width) and reduce them if possible to increase parking supply.

- Establish (in the code) authorization for parking advisory committees for specific areas where parking issues are controversial. Provide for the appointment of a cross section of stakeholders, including businesses and residents. Charter the committee to advise on parking studies and on potential changes to parking ordinances.

2. Major Modifications

- Undertake a comprehensive revision of the parking ordinance. Some specific revisions might include:

  - Revise the tables of parking supply minimums, reducing them wherever possible to reflect context, transportation options, and land use mix.

  - Develop a system of shared parking credits, either as a set percentage in connection with form-based codes or based on the land use mix in connection with zoning.

- Create parking overlay districts for downtowns and mixed-use centers, and write provisions for future additional overlay districts.

- Unbundle parking from residential development in districts with higher densities and a mix of uses.

- Allow off-site parking in dense retail districts and set limits for its distance from development sites.

- Develop provisions to govern joint parking (i.e., parking allowed through contracts or leases with other businesses or landowners) to ensure that parking supply commitments made in connection with development approval are honored and maintained over time.

- Allow some credit for on-street parking supply in retail districts. Allow for substitution of a form-based code in certain zone districts to simplify and eliminate the need for more detailed parking regulations.

- Overhaul the parking enforcement system. Improve enforcement of parking time limits by acquiring hand-held computers for issuing tickets (replacing a system of chalking tires). Revise the parking overtime ordinance to provide escalating fines for scofflaws (repeat offenders) and set fines at levels that deter abuse. Increase enforcement levels so that probability of being ticketed for overtime parking approaches certainty. Evaluate parking supply in and around parking overlay districts and identify parking supply to be available for commuter parking use. Develop a Residential Parking Permit (RPP) system to help protect neighborhoods impacted by overflow parking resulting from increased parking enforcement.

3. Wholesale Changes

- Work with the local or regional transit agency to develop a commuter transit pass that is bundled with a parking permit in parking districts and paid for with proceeds from the district's revenues, including tax revenues. Use this "universal pass" to increase transit patronage while managing commuter parking demand.

- Institute paid parking for public parking supply in parking districts. Start with off-street, publicly owned parking. Pay kiosks for on-street parking can reduce streetscape impacts such as visual clutter from individual parking meters, are more efficient, and are more convenient for customers.

PRACTICE POINTERS

- Implement design standards for parking structures.

- Tailor parking standards for infill areas as opposed to greenfield sites (e.g., fewer, smaller spaces in infill).

- Provide priority parking for hybrid or alternative-fuel vehicles to encourage use of these vehicles.

- Consider requiring a portion of the parking lot to be constructed of pervious materials.

EXAMPLES AND REFERENCES

- Shoup, D. *The High Cost of Free Parking*. Planners Press, American Planning Association. 2005. Chapter 20.

- Metropolitan Transportation Commission. *Developing Parking Policies to Support Smart Growth in Local Jurisdictions: Best Practices*. April 2007. pp. 14-18. http://www.mtc.ca.gov/planning/smart_growth/parking_study/April07/bestpractice_042307.pdf.

- U.S. EPA. *Parking Spaces / Community Places: Finding the Balance Through Smart Growth Solutions*. February 2006. EPA 231-K-06-001. http://www.epa.gov/smartgrowth/parking.htm.

- Maryland Governor's Office of Smart Growth. *Driving Urban Environments: Smart Growth Parking Best Practices*. March 2006. pp. 5-6. http://www.smartgrowth.state.md.us/pdf/Final%20Parking%20Paper.pdf.

- Litman, T. *Parking Management: Strategies, Evaluation, and Planning*. Victoria Transport Policy Institute. November 2008. p. 15. http://www.vtpi.org/park_man.pdf.

- Fitzgerald & Halliday, Inc. *Northwest Connecticut Parking Study - Phase II: Model Zoning Regulations for Parking for Northwestern Connecticut*. Northwestern Connecticut Council of Governments and Litchfield Hills Council of Elected Officials. September 2003. http://www.fhiplan.com/PDF/NW%20Parking%20Study/NW%20Connecticut%20Parking%20Study%20Phase%202.pdf.

- Forinash, C. et al. "Smart Growth Alternatives to Minimum Parking Requirements." Proceedings from the 2nd Urban Street Symposium. July 28-30, 2003. http://www.urbanstreet.info/.

- Victoria Transport Policy Institute. "Parking Maximums." *TDM Encyclopedia*. http://www.vtpi.org/tdm/tdm28.htm#_Toc128220478. Accessed April 12, 2009.

5 INCREASE DENSITY AND INTENSITY IN CENTERS

INTRODUCTION

Density is probably the most discussed and least understood concept in urban planning. Residents and elected officials routinely see the amount of development (e.g., the number of dwelling units, the square footage of commercial space) allowed on a site as one of the most important consideration in local planning. "Too much" density is often seen as the cause of traffic congestion, ugly buildings, loss of green space, crime, and many other ills. However, increasing the average density of infill, redevelopment, and greenfield projects is crucial to improving the quality of life in the community. Higher density is important to protecting open space and supporting transportation options like transit, walking, and biking. Furthermore, EPA research[5] shows that higher densities may better protect water quality—especially at the lot and watershed levels.

As a development center, the Ballston neighborhood of Arlington, Virginia has been designated to accommodate additional growth.

---

5   U.S. EPA. *Protecting Water Resources Through Higher-Density Development.* 2006. EPA 231-R-06-001.

Much of what people dislike about density is in reality the result of development patterns that help to increase congestion on arterials, single-use areas that emphasize driving to get to destinations, and dense developments that are poorly designed. And, unfortunately, many people associate density with poorly managed rental or affordable housing developments. Fear of lower property values is often an underlying concern of residents when discussing higher density developments.

Density itself does not determine the quality of development. Many high-density areas, in fact, are the most desirable areas in a region, such as Dupont Circle in Washington, D.C., and the Chicago suburb of Oak Park, Illinois. These areas are attractive because the density is well designed, with appealing streetscapes, mixture of uses, site planning, and building design. Despite the multiple benefits that can be derived from projects with higher densities, gaining political approval for higher density projects is often difficult and controversial.

Desire for privacy, feeling crowded, fear of crime, parking, and compatibility with the character of the community are often the issues that residents cite as concerns with more dense developments. Identifying techniques and requirements to ensure that higher density projects are compatible with existing neighborhoods will help respond to these concerns.

RESPONSE TO THE PROBLEM

The concept of density requires ample discussion and education to allay misconceptions and correct misunderstandings about its purpose and benefits. Increased density creates the customer base needed for transit, retail, and amenities residents want. Residents of less dense communities may ask, "Why can't we have the amenities that that community has?" Often, the answer is that the other community is denser. The benefits and resources discussed in this section provide the foundation for a complete community, one that needs increased density to thrive.

Communities need to address density in a comprehensive manner rather than project by project. There are a number of strategies and tools that communities may use to decide which parts of their community should be densest. Through the comprehensive or general plan process, the community should target areas that have the character and infrastructure

to support higher density development. Communities should ensure that higher density developments go into mixed-use areas that will allow walking and biking to shops and services, which reduces driving and can minimize parking requirements. Lastly, communities should focus much of their higher density where it can be served conveniently by bus or rail transit, which will also reduce the need to drive and provide other environmental benefits.

These policies can be implemented through new mixed-use or transit-oriented development (TOD) districts, changes in zoning designations, or modifying zoning to allow greater density in existing districts. Other strategies include creating new compatibility standards and design guidelines to improve transitions between higher density development and low-density neighborhoods.

EXPECTED BENEFITS

- Less pressure to expand development to outlying areas, thus protecting agricultural lands, natural open space, bodies of water, or sensitive habitat.

- Buildings and developments that use less energy, less land, and typically less materials. Because of the more efficient buildings and the transportation options that reduce the need to drive, residents generate fewer greenhouse gases per capita.

- More diverse communities with more opportunities for affordable housing, particularly in areas that have high land values and scarce development sites.

- More effective transit service. In lower density neighborhoods, seven to eight units per acre is the minimum density necessary to support transit service.[6]

- Support for local shops and services that rely on customers who can walk or bike from surrounding neighborhoods.

---

6   Dittmar, H. and Ohland, G. *The New Transit Town.* 2003.

STEPS TO IMPLEMENTATION

1. Modest Adjustments

- Set minimum (as opposed to maximum) densities in general or comprehensive plans and zoning districts. This tool helps creates neighborhoods that are close-knit and vibrant and helps achieve benchmarks for citywide housing policies and goals.

- Designate locations for higher density development centers in comprehensive plans.

- Create activity center districts with higher densities, increased heights and FAR, and reduced parking requirements. This can be done by creating specific zones, modifying existing zones, or creating a new overlay district that allows selective modification of existing zoning regulations in an already zoned area without changing all of the zoning of a parcel.

2. Major Modifications

- Tailor development standards (e.g., height limits and FAR, parking requirements, and open space and landscaping regulations) to accommodate denser developments. Urban-style projects should not be evaluated based on low-density development standards.

- Rezone areas designated as activity centers based on comprehensive plans to increase density, as opposed to using case-by-case rezoning.

3. Wholesale Changes

- Use a redevelopment agency to purchase difficult-to-obtain or critical parcels. This is particularly effective with areas such as corridors, which often have smaller parcels that require aggregation to allow higher density development.

- Establish minimum densities or intensities in community or regional mixed-use centers and transit-oriented developments.

- Use height, placement, coverage and perviousness requirements, rather than FAR, to regulate structured parking. For example, do not count structured parking toward FAR if it is screened from view with retail, residential or office structures, or is constructed above the ground floor of a structure.

- Parking can be a costly component of development. Parking may be reduced as part of a TOD or a mixed-use, high-density district. Parking may also be "unbundled" from the residential units, which allows residents to choose not to purchase parking. (See Essential Fix No. 4.)

US Environmental Protection Agency

The Back Bay in Boston, Massachusetts serves as a center for commerce, housing and other activities. The intensity of resources here minimizes pressure to develop elsewhere because of available infrastructure and services.

♻EPA United States Environmental Protection Agency

- Set parking maximums rather than minimums to discourage too much parking supply for a development. This will allow higher density development, as parking often limits a project's overall density.

PRACTICE POINTERS

- Density is context sensitive; different levels of density will be appropriate in different places.

- Adopt site and building design standards for higher density projects to ensure high-quality, attractive development.

- Consider offering density bonuses and flexible zoning standards to encourage construction of affordable housing. Many jurisdictions have developed density bonuses, as well as allowable concessions or variances for specific regulations, as an incentive for affordable, senior, or disabled housing.

- Designating a buildable envelope rather than specifying density allows flexibility in the number of units, which creates greater density while controlling variables such as height and setbacks.

- Adopt transition/compatibility standards (e.g., building setbacks, open space, landscaping) to ensure that higher density projects in activity centers are compatible with surrounding neighborhoods.

EXAMPLES AND REFERENCES

- U.S. EPA. *Protecting Water Resources with Higher-Density Development*. January 2006. EPA 231-R-06-001. pp. 44-51. http://www.epa.gov/smartgrowth/water_density.htm.

- State of Georgia. "Minimum Density Zoning." Georgia Quality Growth Toolkit. http://www.dca.state.ga.us/intra_nonpub/Toolkit/Guides/MinDensZning.pdf. Accessed June 30, 2009.

- Edelman, M. "Increasing Development Density to Reduce Urban Sprawl." Iowa State University Extension Service. 1998. http://www.extension.iastate.edu/newsrel/1998/dec98/dec9810.html.

- Coupland, A. *Reclaiming the City: Mixed Use Development*. Routledge. November 1996. p. 35.

- Williams, K. and Seggerman, K. *Model Regulations and Plan Amendments For Multimodal Transportation Districts*. Florida Department of Transportation. April 2004. http://www.dot.state.fl.us/planning/systems/sm/los/pdfs/MMTDregs.pdf.

- Oregon Transportation and Growth Management Program. *Commercial and Mixed-Use Development Code Handbook*. October 2001. pp. 40-43. http://egov.oregon.gov/LCD/docs/publications/commmixedusecode.pdf.

- City of Colorado Springs, Colorado. *Mixed Use Development Design Manual*. March 2004. pp. 56-64 http://permits.springsgov.com/units/planning/Currentproj/CompPlan/MixedUseDev/IV-%20E.pdf.

- Institute for Urban and Regional Development. "Relations between Affordable Housing Development and Property Values." Working Paper 599. University of California, Berkeley. May 1993. http://www.hcd.ca.gov/hpd/prop_value.pdf. Accessed August 27, 2009.

- California Housing Law Project. "SB 1818 – Density Bonus." Fact sheet. 2004. http://www.housingadvocates.org/facts/1818.pdf.

- Shoup, D. *The High Cost of Free Parking*. Planners Press, American Planning Association. 2005. Chapter 20.

 # MODERNIZE STREET STANDARDS

INTRODUCTION

For several decades, municipal decisions about the size and design of streets have been based primarily on traffic capacity considerations. This narrow focus overlooks the fundamental role that streets play in shaping neighborhoods and communities. Streets are an important use of land. The design of streets influences the character, value, and use of abutting properties, as well as the health and vitality of surrounding neighborhoods. Street design also determines whether the area will be walkable, whether certain types of retail will be viable, and whether the urban landscape will be attractive and comfortable or stark and utilitarian. These impacts, in turn, affect land values (and associated tax receipts) and overall economic strength and resiliency. The character of streets can discourage or encourage redevelopment, hasten or reverse urban flight, and add or subtract value from abutting property. These are obviously important policy considerations for any municipality.

Street design also affects environmental factors, including the volume of stormwater runoff, the water quality of that runoff, and the magnitude of the urban heat island effect. Street trees are particularly important: they remove carbon dioxide and certain pollutants from the air; they intercept and absorb rain before it reaches the street; they shade the landscape, reducing ambient air temperatures in warm months; they add aesthetic value to neighborhoods; and they slow traffic, improving public safety.

Cities and towns have tended to make planning and design decisions about streets one project at a time and based on a limited perspective of specific sections of specific streets. This narrow perspective ignores the fact that transportation systems are comprised of networks of facilities. The macro-scale characteristics of *networks* are more important than the micro-scale design of specific street sections in determining how well a local transportation system functions (including how much capacity the system has).

This conventional project-by-project perspective has resulted in poorly connected networks of oversized streets, rather than well-connected networks of smaller streets. The resulting connectivity problems have been exacerbated by the national trend, beginning in the 1920s, of letting developers make network layout and connectivity decisions for streets built as part of their subdivisions and commercial sites. The inevitable outcomes have been poor connectivity, inconvenient circulation, and over-crowded arterials. These outcomes, in turn, have been detrimental to emergency service response, access to existing businesses, and neighborhood walkability.

The issues around street design and network connectivity have been further compounded by oversimplified and unsupported theories about traffic safety. In recent years, transportation engineering analysis has shown that street width; the size, proximity, and orientation of buildings and street trees; the configuration of intersections; and the presence of on-street parking all have significant effects on the speed and attentiveness of drivers. Designed properly, these elements can reduce both accident frequency and accident severity.

Clearly, there is a need for communities to update their approach to planning, designing, and building streets and street networks.

This view of University Boulevard in Palo Alto, California includes amenities for cars and bikes.

RESPONSE TO THE PROBLEM

Generally, cities have addressed street design issues through subdivision regulations rather than zoning ordinances, although that varies depending on the local regulatory structure. Form-based codes can provide a foundation for street design and, to a lesser extent, for connectivity, but additional design details and procedural requirements will be needed. The primary techniques that cities and towns are implementing to improve street design include:

- **Complete Streets** – Streets should be designed to serve all modes of travel equally well—pedestrians, bicycles, personal vehicles, and transit.

- **Narrow Local Streets** – Local streets (streets that primarily provide access to abutting properties, as opposed to streets that primarily serve pass-through traffic) should be no wider than absolutely necessary.

- **Context-Sensitive Thoroughfares** – Arterial and collector thoroughfares should be designed to fit the character of abutting lands and surrounding neighborhoods and should not be overly wide or designed to encourage inappropriate vehicular speeds.

- **Pedestrian-Oriented Environments** – Streets should be walkable—safe, attractive, and convenient for pedestrians, including people walking for utilitarian purposes as well as people strolling and exercising.

- **Universal Design** – Pedestrian facilities should be designed to be convenient and safe for a wide variety of people, including persons with disabilities, elderly people and children, people pushing strollers, and strong, fit pedestrians walking quickly.

- **Green Streets** – Streets can be designed with features that manage stormwater and protect water quality by reducing the volume of water that flows directly to streams and rivers; using a street tree canopy to intercept rain, provide shade to help cool the street, and improve air quality; and serving as a visible element of a system of green infrastructure that is incorporated into the community.

- **On-Street Parking** – On-street parking is not only a convenient way to add value to properties in mixed-use districts. It can also be a design strategy to make streets safer and more appealing for pedestrians.

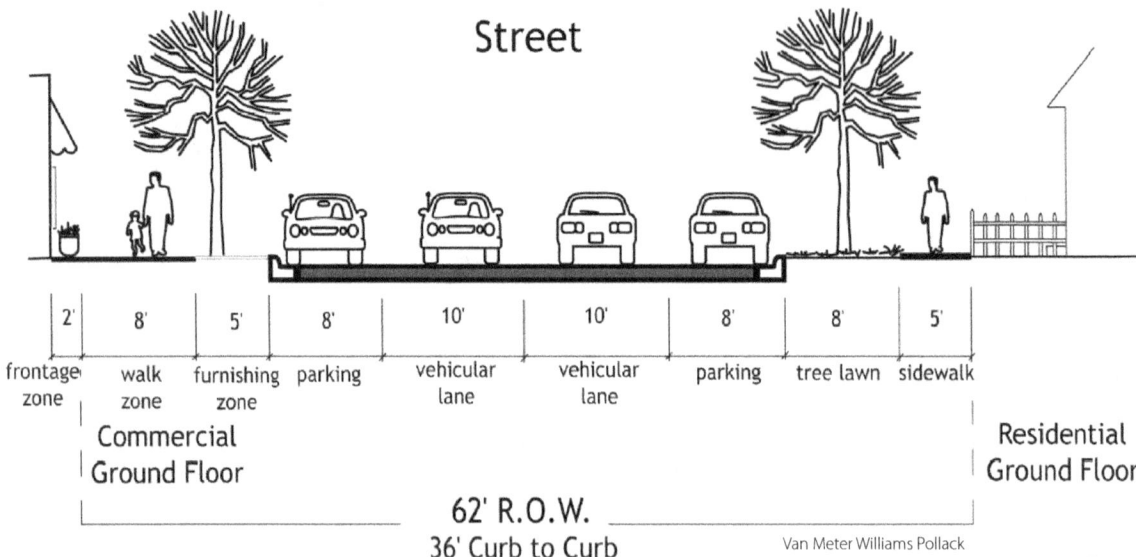

Street

2'	8'	5'	8'	10'	10'	8'	8'	5'
frontage zone	walk zone	furnishing zone	parking	vehicular lane	vehicular lane	parking	tree lawn	sidewalk

Commercial Ground Floor

Residential Ground Floor

62' R.O.W.
36' Curb to Curb

Van Meter Williams Pollack

This street section show the typical array of uses for a right of way including pedestrians and automobiles.

Many communities, along with state departments of transportation, are addressing network connectivity issues by changing their land development codes and subdivision regulations to require minimum connectivity in new development and in redevelopment. To be effective, these standards must address both external connectivity (how well connected a development is with the larger street network) and internal connectivity (how well the land uses in the development are connected with each other). The most commonly used connectivity regulations establish standards for:

- Maximum block length and circumference or block area;

- Minimum intersections per linear mile of roadway or per square mile of area; and

- Connectivity Index (the number of street links divided by the number of intersections).

EXPECTED BENEFITS

- Improved safety for drivers, pedestrians, and bicyclists.

- Reduced environmental footprint, including less stormwater runoff, less of a heat island effect, and less land consumed.

- More walking and biking with attendant health benefits.

- Value added to abutting properties and surrounding neighborhoods.

- Increased tax base and tax revenues.

- A more attractive city or town with more economic vitality and resiliency.

- A more flexible, adaptive network to help avoid congestion.

- Improved emergency response and emergency evacuation capability.

- Reduced street maintenance costs.

- Allowing people to drive less with no reduction in mobility.

STEPS TO IMPLEMENTATION

1. Modest Adjustments

- Revise the local street design standards to add a "road diet" cross section for appropriate streets that currently have four general purpose lanes with no on-street parking, no bike lanes, inadequate pedestrian space, or any combination of these deficiencies. Set criteria for conversion to three lanes (two general purpose lanes and a two-way left turn lane) with either bike lanes or on-street parking and improved pedestrian amenities.

- Update the local street design standards to include universal design criteria for pedestrian curb ramps, crosswalks,

and curb extensions. Create overlay design criteria for Safe Routes to School programs, transit corridors, downtowns, and other priority pedestrian areas.

- Update design standards governing provision of street trees to increase the city's street canopy as new streets are built and as existing streets undergo major renovation. Clearly and permanently resolve issues of cost responsibility for maintenance of street trees. Ensure that standards are realistic for the local climate, specifying appropriate tree species and appropriate designs to contain tree root structures.

- Adopt a policy governing provision of bike lanes on arterials and collectors as streets are built and as existing streets undergo major renovation. Set standards for deciding which streets will have on-street lanes, taking into account spacing of facilities, speed of traffic, availability of right of way, and other practical matters. This policy will be most effective if it is based on a local bicycle system plan that sets system objectives, defines facility types, and sets connectivity standards.

- Begin developing and testing stormwater management designs such as rain gardens, bio-swales, and other techniques in preparation for development of green streets standards and policies.

2. Major Modifications

- Because streets are integral to community form and character, the best way to set the stage for improvements in street design and street network connectivity is to embed street design principles in the comprehensive plan or community master plan. In states and regions with growth management or environmental requirements governing preparation of local plans, this will be a necessary step prior to the measures described below. In most places, the planning foundation should take the form of a multimodal transportation master plan or a multimodal transportation element in the comprehensive plan.

- Revise the street classification system to create a "multimodal corridor" designation. This can also be handled as an overlay requirement without changing the underlying functional classification system. Use the multimodal corridor designation to apply complete streets principles (design for all modes) in specific corridors. A network of multimodal corridors based on local transit routes and on a bicycle system plan can guide both development review and prioritization of projects in a capital improvements program. This should be an interim step toward implementation of complete streets requirements community-wide.

- Revise street design standards to add "narrow local streets" categories. Create design templates for residential and commercial streets that are narrower than currently allowed.

- Set minimum internal connectivity standards for new subdivisions based on maximum block length, block size, intersections per square mile, or a Connectivity Index.

- Create a policy or update existing requirements to prevent any street abandonment or closure that would reduce the connectivity of the street network.

3. Wholesale Changes

- The need for a planning foundation applies to measures in this section as well. All of the measures described below should be based on an adopted multimodal transportation master plan or multimodal transportation element in the comprehensive plan.

- Overhaul the street design standards with the objective of reducing the future environmental footprint of streets. Incorporate complete streets provisions and green streets principles. Adopt narrower lanes, narrower rights of way, and reduced-lane cross sections.

- Reintroduce public alleys into the local transportation system. Create standards allowing and guiding provision of alleys in subdivisions and requiring them in large commercial projects. Add alley templates to the local street design standards.

- Set minimum internal and external connectivity standards to be applied to all new subdivisions and large commercial projects and to guide local public works decision-making relative to the capital improvements program.

- Update the code to significantly increase the amount of on-street parking in commercial and mixed-use districts and on residential streets.

PRACTICE POINTERS

- Involve emergency service providers and the public works and other departments early in comprehensive planning and before code revisions are drafted. Narrower lanes and reduced-lane cross sections can be controversial, and city councils may be unwilling to override a fire chief's concerns about these issues. In many cases, coordination and cooperation between local departments have overcome such obstacles.

- In many states, at least some degree of state guidance applies to local street design standards. And in virtually any municipality, some important streets will be under state jurisdiction (e.g., state routes). For these reasons, early and continuing coordination with the state department of transportation is critical to the success of most of the measures outlined above.

- Look for opportunities for cost savings and other benefits associated with narrower street standards, including reduced stormwater volume, reduced snow removal and other maintenance costs, and other savings.

EXAMPLES AND REFERENCES

- Handy, S., Paterson, R., and Butler, K. *Planning for Street Connectivity: Getting from Here to There.* Planning Advisory Service Report Number 515. American Planning Association. May 2003. pp. 12-15. http://www.planning.org/apastore/search/default.aspx?p=2426

- Institute of Transportation Engineers. *Context Sensitive Solutions in Designing Major Urban Thoroughfares for Walkable Communities: An ITE Proposed Recommended Practice.* May 2005. http://www.ite.org/emodules/scriptcontent/orders/ProductDetail.cfm?pc=RP-036. (Note: this document is being updated and is expected to be issued as a final recommended practice in late 2009.)

- Larimer County, Colorado. *Larimer County Urban Area Street Standards.* April 2007. http://www.larimer.org/engineering/gmardstds/UrbanSt.htm. Accessed June 25, 2009.

- City of Charlotte, North Carolina. *Urban Street Design Guidelines.* October 2007. http://www.charmeck.org/Departments/Transportation/Urban+Street+Design+Guidelines.htm. Accessed June 25, 2009.

- Williams, K. and Seggerman, K. *Model Regulations and Plan Amendments For Multimodal Transportation Districts.* Florida Department of Transportation. April 2004. http://www.dot.state.fl.us/planning/systems/sm/los/pdfs/MMT-Dregs.pdf.

- National Complete Streets Coalition. http://www.complet-estreets.org.

- City of Roanoke, Virginia. *Street Design Guidelines.* July 2007. http://www.roanokeva.gov/85256A8D0062AF37/CurrentBaseLink/7C223BF47CE37256852575F2006CEDF8/$File/STREET_DESIGN_GUIDELINES.pdf. Accessed June 14, 2009.

- Duany Plater-Zyberk & Company. *SmartCode, Version 9.2.* February 2009. http://www.smartcodecentral.com/smartfilesv9_2.html.

- American Association of State Highway and Transportation Officials. *Guidelines for Geometric Design of Very Low-Volume Local Roads (ADT ≤ 400).* 1st Edition. June 2001. https://bookstore.transportation.org/imageview.aspx?id=450&DB=3.

- Neighborhood Streets Project Stakeholders. *Neighborhood Street Design Guidelines: An Oregon Guide for Reducing Street Widths.* Oregon Transportation and Growth Management Program. November 2000. http://www.oregon.gov/LCD/docs/publications/neighstreet.pdf.

- Mozer, D. "Planning: Bicycle and Pedestrian Friendly Land Use Codes." International Bicycle Fund. April 2007. http://www.ibike.org/engineering/landuse.htm. Accessed September 13, 2009.

- Metro Regional Government (Portland, Oregon). *Green Streets: Innovative Solutions for Stormwater and Stream Crossings.* June 2002. http://www.oregonmetro.gov/index.cfm/go/by.web/id=26335.

- City of Boulder, Colorado. "Multimodal Corridors." April 2006. http://www.bouldercolorado.gov/index.php?option=com_content&task=view&id=355&Itemid=1624. Accessed June 12, 2009.

ENACT STANDARDS TO FOSTER WALKABLE PLACES

INTRODUCTION

In smart growth communities, people are able to walk comfortably and safely to work, school, parks, stores, and other destinations. Current codes in many communities, however, result in places that prevent or discourage walking by imposing low-density design (see Essential Fix No. 2), including overly wide streets and landscapes designed for cars instead of people (see Essential Fix No. 6). In such places, the pedestrian realm is treated as an afterthought—the space left over between the edge of the street and the buildings and parking lots. One significant challenge to developing a walkable community is the lack of design standards or performance measures for walkability, like those that guide other kinds of transportation planning and design. Thus many communities are not in a position to guide private development and public works investments to build good pedestrian accommodation into development and redevelopment, and they do not have programs or provisions to repair older, pedestrian-hostile areas. The magnitude of this need has been highlighted in recent years both by the number of pedestrian injuries and fatalities and by the health effects that less physical activity—which is often a direct result of urban design—have had on the U.S. population.

RESPONSE TO THE PROBLEM

The two primary elements to be addressed through codes are design standards for facilities, including public works facilities built by and for the city (e.g., streets and sidewalks), and requirements for private development and redevelopment projects. Communities usually regulate facility design through design standards adopted as ordinances or as administrative rules. In addition to guiding the planning and design decisions for municipal facilities, these design requirements may

Van Meter Williams Pollack

Pearl Street in Boulder, Colorado shows the street view of how wide sidewalks can contribute to a pleasant walkable experience.

be applied to private projects in part through the zoning approval process and in part through subdivision regulations. In some communities, form-based codes are used not only to guide the design of streets and sidewalks, but also to create a connection between all elements of the built environment. Communities may also use level of service[7] standards to ensure that development and redevelopment projects meet minimum criteria for walkability. Finally, communities may adopt Safe Routes to School program planning and design criteria and may designate pedestrian districts or zones in special areas (e.g., in downtowns, around schools, near colleges and universities).

The fountain and plaza located at the entrance of a bookstore act as a central gathering and meeting space in Bethasda Row.

EXPECTED BENEFITS

- Safer communities with fewer pedestrian injuries and deaths from vehicle collisions.

- Healthier people because of more opportunities to walk or bike.

- More economically viable places, stabilized property values, and reduced retail leakage (where potential patrons go elsewhere, perhaps due to a lack of safe walking conditions).

- Increased transit ridership because of better pedestrian access to transit.

- Reduced parking demand in commercial areas due to "park once" strategy.

- Reduced driving as short trips are made by walking rather than driving.

- Reduced per capita emissions of criteria air pollutants[8] and greenhouse gases resulting from reduced driving.

STEPS TO IMPLEMENTATION

1. Modest Adjustments

- Develop or revise street and street crossing design standards to improve pedestrian safety, convenience, and comfort, both as a part of routine public works projects and as a part of ongoing development and redevelopment.

- Adopt standards to incorporate trees and other shade structures into the pedestrian realm, especially in mixed-use districts, addressing maintenance and irrigation as well as landowner responsibilities.

- Prepare and implement a Safe Routes to School program, taking advantage of federal funding and a national database of successful examples.

2. Major Modifications

- Designate one or more pedestrian districts (keep the initial number small) where the community will focus its efforts to make walking safer and more pleasant. Develop

---

7   Level of service is a measure of effectiveness by which traffic engineers determine the quality of service of elements of transportation.

8   Criteria pollutants are monoxide, lead, nitrogen dioxide, ozone, particulate matter, and sulfur dioxide and are regulated by EPA under the Clean Air Act.

United States Environmental Protection Agency

a zoning overlay district to make targeted changes to the underlying zoning categories to reallocate street cross sections, regulate building setbacks, and so forth. Prioritize capital improvement funding to pedestrian facility needs in the zoning overlay district. Build upon success by designating additional pedestrian districts once the program has solid achievements to show in the initial district(s).

- Establish pedestrian level of service and connectivity requirements for all development and redevelopment projects of more than two acres. Include minimum pedestrian connectivity within developments and with adjacent developments.

- Adopt pedestrian environment standards for mixed-use districts to improve pedestrian safety, comfort, and convenience, including requirements for on-street parking, build-to lines, minimum façade transparency, building entrance spacing, canopies, and similar pedestrian-friendly elements.

3. Wholesale Changes

- Prepare and adopt a pedestrian circulation element in the comprehensive plan or in a separate transportation master plan. Develop a prioritized multi-year pedestrian capital improvements plan to implement the circulation element.

- Require major developments to include pedestrian circulation plans as part of application or site plan submittals. Set and apply minimum connectivity standards and level of service criteria.

- Revise subdivision and zoning development standards to require sidewalks on both sides of streets in all developments.

- Require walkways in parking lots larger than 1 acre or 200 feet wide, linking perimeter sidewalks to primary building entrances.

PRACTICE POINTERS

- Communities often adopt plans calling for the entire community to be "pedestrian friendly." This often turns out to be more a slogan than a policy. Virtually any community in the United States today has vast areas of landscape with poor pedestrian accommodation, and fixing these areas will take many years of investment and careful regulation. Communities should implement regulations that prevent new development of areas with inadequate pedestrian accommodation and adopt standards that prevent construction of any new streets with inadequate provisions for pedestrians. Public investment to retrofit and improve sidewalks, crosswalks, grade separations, and other facilities should go initially to school zones and routes, downtowns and other mixed-use districts, transit corridors, and other areas where a significant pedestrian presence is expected or desired.

- Involve a wide range of stakeholders and city departments (e.g., fire, police, public works) throughout any pedestrian circulation planning process.

- One of the most important characteristics of public streets affecting pedestrian environments is the speed of vehicular traffic. Speeds above 30 mph make sidewalks less pleasant and street crossings more dangerous and difficult.

- The most critical link in any pedestrian network is the availability of safe, appropriately spaced street crossings, especially crossings of arterial streets. Communities need good policies for location, frequency, and design of street crossings, and they must invest in safe, well-designed crossings if they want to develop functional, active pedestrian districts.

- On-street parking is an important pedestrian feature that protects walkers by separating sidewalks from moving traffic. On-street parking also makes it easier for people to walk to their destinations.

- Cities must stay current with universal design requirements that ensure sidewalks, trails, crosswalks, parking lots, building entrances, and other features of the built environment are fully accessible to people with physical disabilities and other physical challenges. The national Americans with Disabilities Act outlines specific regulatory requirements, which are expanded and updated frequently.

EXAMPLES AND REFERENCES

- Florida Department of Transportation. *Multimodal Transportation Districts and Areawide Quality of Service Handbook.* November 2003. p. 26. http://www.dot.state.fl.us/planning/systems/sm/los/pdfs/MMTDQOS.pdf.

- National Complete Streets Coalition. http://www.completestreets.org.

- Dixon, L. "Bicycle and Pedestrian Level-of-Service Performance Measures and Standards for Congestion Management Systems." *Transportation Research Record 1538.* 1996. http://www.enhancements.org/download/trb/1538-001.PDF.

- Landis, B. et al. *Modeling the Roadside Walking Environment, A Pedestrian Level of Service.* Transportation Research Board Paper No. 01-0511. 2001. http://www.dot.state.fl.us/planning/systems/sm/los/pdfs/pedlos.pdf.

- U.S. Green Building Council. LEED for Neighborhood Development Rating System Credit for Walkable Streets (Neighborhood Pattern and Design, Credit 7, in pilot version). http://www.usgbc.org/DisplayPage.aspx?CMSPageID=148. Accessed June 20, 2009.

- Duany Plater-Zyberk & Company. *SmartCode, Version 9.2.* February 2009. http://www.smartcodecentral.com/smartfilesv9_2.html.

- Ewing, R. *Pedestrian and Transit-Friendly Design: A Primer for Smart Growth.* International City/County Management Association and Smart Growth Network. 1999. http://www.epa.gov/smartgrowth/pdf/ptfd_primer.pdf.

- Federal Highway Administration. *Designing Sidewalks and Trails for Access: Part I of II: Review of Existing Guidelines and Practices.* 1999. http://www.fhwa.dot.gov/environment/sidewalks/index.htm.

- Federal Highway Administration. *Designing Sidewalks and Trails for Access: Part II of II: Best Practices Design Guide.* 2001. http://www.fhwa.dot.gov/environment/sidewalk2/index.htm.

- City of Redmond, Washington. "Pedestrian Program Plan." *Transportation Master Plan.* November 2005. http://www.redmond.gov/connectingredmond/policiesplans/tmpprojectdocs.asp.

- National Center for Safe Routes to School. http://www.saferoutesinfo.org.

- City of Seatac, Washington. Pedestrian Overlay District. *Seatac Zoning Code.* November 2002. http://mrsc.org/mc/seatac/stac1528.html. Accessed May 5, 2009.

- Cleveland Neighborhood Development Coalition. Pedestrian Retail Overlay (PRO) District. http://www.cndc2.org/prod.html. Accessed May 5, 2009.

- Leaf, W.A. and Preusser, D.F. "Literature Review on Vehicle Travel Speeds and Pedestrian Injuries." U.S. Department of Transportation. DOT HS 809 021. October 1999. http://www.nhtsa.dot.gov/people/injury/research/pub/hs809012.html.

- Federal Highway Administration "Safe Routes to School: Program Legislation – SAFETEA-LU, Sec. 1404." http://safety.fhwa.dot.gov/saferoutes/overview/legislation.cfm#sec1404. Accessed May 5, 2009.

DESIGNATE AND SUPPORT PREFERRED GROWTH AREAS AND DEVELOPMENT SITES

INTRODUCTION

For many decades, most municipalities have handled land development and growth reactively. Zoning changes have been initiated primarily by landowners and developers. Developers have often selected development locations that did not follow city comprehensive plans. Subdivision and property assembly have been undertaken by landowners and developers with specific development projects in mind. There is often a financial incentive for developers to develop peripheral sites rather than redeveloping infill sites. However, communities can better control the development they get by focusing their resources to catalyze redevelopment in desired areas.

Planning land uses and development intensities in preferred growth areas and development sites generates several benefits. It encourages and facilitates redevelopment and infill, supports transit, and guides new development to appropriate areas with ready access to existing infrastructure. Local governments need to play a more active role in selecting areas where new growth makes the most sense. They need to reinforce those choices by revising their development codes and capital improvement plans to make these areas more attractive to the development community than other, less appropriate areas. This more focused approach to development can benefit both individual landowners and the entire community.

A palm tree-lined pedestrian plaza leads to the entrance of the largest apartment buildings at the center of Mizner Park in Florida. Higher densities in this existing development enable greenfields to be preserved.

RESPONSE TO THE PROBLEM

Municipalities need to be proactive about determining where and to what extent they will grow. This planning can provide government officials with the justification to say "no" to development proposals that are not in the community's best interests and are inconsistent with the community plan. Even in communities that cannot keep up with infrastructure needs, many local governments believe there is benefit in encouraging more development. But to be effective on behalf of current residents and thoughtful about the needs of future residents, cities need to designate where growth will occur, then rezone, change codes, and alter utility and infrastructure provisions to accommodate that growth.

To focus development where it makes the most sense, a community needs a detailed plan. This plan should include comprehensive subdivision regulations and street mapping, zoning, and design guidelines, as well as an infrastructure plan and a financing or implementation plan. Developing the plan should include a comprehensive stakeholder and public engagement process. The designation of growth areas should be supported by studies and data, such as a fiscal impact analysis or a cost of infrastructure study.

EXPECTED BENEFITS

- Greater predictability for infill proposals that meet the new development standards, and certainty of location and development potential for landowners, developers, and citizens.

- More efficient development review processes. Complete policies on land use and development regulations will help streamline the review process and garner stronger support from the planning commission and/or city council.

- Cost-effective provision of infrastructure. Focusing on and prioritizing infill development will use existing infrastructure efficiently.

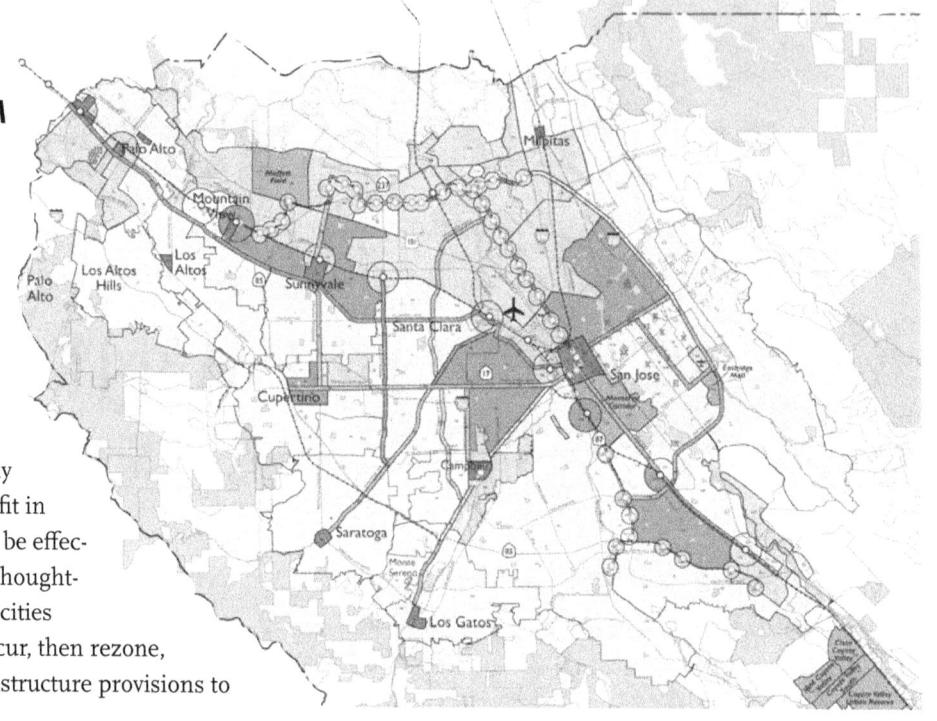

This rendering of Santa Clara, California illustrates how the city has designated preferred growth areas to keep distinctive places intact.

- Preservation of open space and natural resources. Focusing on infill development reduces pressure to expand on a community's periphery or to develop in areas with sensitive habitat or open space.

STEPS TO IMPLEMENTATION

(Note: Steps may be applied differently in infill versus greenfield locations.)

1. Modest Adjustments

- Identify and map preferred growth areas in a comprehensive plan. The plan should include goals and objectives for the various areas.

- Establish utility and transportation capacity plans.

- Change the minimum lot size, requiring smaller parcels to be aggregated or developed in conjunction with larger parcels in a coordinated manner.

- Designate agriculture interim/holding zones in lieu of low-density zoning in areas where the local government would rather not see imminent development.

- Create district or area plans to guide development.
- Vary fees for development based on location, as infill sites usually have lower infrastructure costs than peripheral or greenfield development.

2. Major Modifications

- Enact an adequate public facility ordinance (APFO). An APFO helps ensure that infrastructure for schools, road, sewers, and fire protection exists to accommodate new development.
- Establish a policy that sets criteria for annexation, including the provision of utilities, infrastructure financing, and minimum development thresholds. The policy should also include requirements for developing an annexation plan for the area. (See Essential Fix No. 10 for more on annexation issues.)
- Establish urban service areas or boundaries as part of the overall master facilities plan to help phase development in coordination with infrastructure.

3. Wholesale Changes

- Establish urban service areas or growth boundaries, and support them by zoning areas outside the boundaries for agriculture and other very low-density uses.
- As part of detailed area plans, rezone designated growth areas (e.g., around transit stops or regional activity centers) to allow denser development.

PRACTICE POINTERS

- Coordinate local government capital investment plans to support development in designated growth areas and to discourage it in other areas.
- Adopt a comprehensive plan land use map that depicts preferred development areas and clearly describes the mix of uses, community design principles, and key features desired for each area.
- Coordinate with other local governments in the region to adopt supportive plans and designated growth areas. It is extremely important to coordinate what will happen in the areas between cities so that these community separators can be maintained over time.

- It is also critical to strategically manage the phasing of growth areas. Each town or city needs to find the appropriate strategy for holding growth areas in check until they are prepared for the types of development that the community envisions.
- Communities need to find ways to prioritize development so that key projects can be implemented earlier as catalysts. Often, lower intensity or less complex developments will be attempted first, which sometimes robs critical or desired projects of their market opportunity and thus pushes them off for many years. This is particularly true of retail, which requires residential support and typically will be drawn to automobile-oriented sites before the infill sites the community may desire.

EXAMPLES AND REFERENCES

- Porter, D. "Chapter 3: Managing Community Expansion: Where to Grow." *Managing Growth in America's Communities*. Island Press. November 2007.
- Nolon, J.R. "Chapter 2: Local Land Use Controls That Achieve Smart Growth." *Well Grounded: Using Local Land Use Authority to Achieve Smart Growth*. Environmental Law Institute. July 2001.
- Lancaster County, Pennsylvania. "Designated Rural Area Concept." *Lancaster County Comprehensive Plan*. June 2005. http://www.co.lancaster.pa.us/planning/lib/planning/long_range/growth_management/rural_area_concept_summary.pdf.
- City of Austin, Texas. Smart Growth Initiative. http://www.ci.austin.tx.us/smartgrowth. Accessed June 10, 2009.
- City of Austin, Texas. *Smart Growth Criteria Matrix*. February 2001. http://www.epa.gov/smartgrowth/scorecards/austin_matrix.pdf.
- State of Maryland. *Smart Growth Priority Funding Areas Act of 1997*. http://www.mdp.state.md.us/fundingact.htm. Accessed April 22, 2009.
- City of Boulder, Colorado. B*oulder's Open Space & Mountain Parks: A History*. http://www.bouldercolorado.gov/index.php?option=com_content&task=view&id=1167&Itemid=71. Accessed May 12, 2009.

9 USE GREEN INFRASTRUCTURE TO MANAGE STORMWATER

INTRODUCTION

Many communities across the United States face the challenge of balancing water quality protection with accommodating new growth and development. Conventional development practices cover large areas with impervious surfaces such as roads, driveways, and buildings. Once such development occurs, rainwater cannot infiltrate into the ground. Instead, it runs off the land at much higher levels than would naturally occur. The collective force of this runoff scours streams, erodes stream banks, and carries large quantities of sediment and other pollutants into waterbodies each time it rains. Most municipal stormwater regulations require stormwater management only at the site scale, using pipes, curbs, gutters, and basins. This approach has functioned well to mitigate local flooding but has resulted in degraded waterways and poor water quality at the watershed scale. A conventional approach to managing stormwater at the site scale fails to address the impacts of land use on water quality, particularly:

- Loss of natural land and disruption of water systems;
- Increased impervious surface area; and
- Increased stormwater runoff volumes.

Many local ordinances besides stormwater regulations pose barriers to better stormwater management and watershed protection. Communities must also look beyond the site scale and consider the impacts of where and how development occurs across neighborhoods and watersheds.

This picture illustrates site level green infrastructure practices such as landscaped swales to capture runoff.

RESPONSE TO THE PROBLEM

Communities are recognizing that the water quality impacts of development need to be managed at a variety of scales, including the municipal, neighborhood, and site levels. Green infrastructure uses natural and built systems at all three scales to protect water quality.

At the regional or watershed scale, green infrastructure is the interconnected network of preserved or restored natural lands and waters that provide essential environmental functions. At the community or neighborhood scale, green infrastructure incorporates planning and design approaches such as compact, mixed-use development; parking reductions; and street trees and other vegetation that reduce impervious surfaces and make communities more attractive. At the site scale, green infrastructure mimics natural systems by holding stormwater in rain gardens or swales to allow it to absorb into the ground (infiltration), using trees and other vegetation to convert it to water vapor (evapotranspiration), and using rain barrels or cisterns to capture stormwater for reuse.

Changing codes to support green infrastructure at all three scales protects water quality while creating many other environmental, community, and economic benefits. Local governments can incorporate green infrastructure by adopting plans, removing barriers, enacting regulations, and creating incentives for green infrastructure on both public lands and private property. Certain local policies, such as landscaping and parking requirements or street design criteria, can complement strong stormwater standards and make it easier for developers to simultaneously meet multiple requirements.

Communities can incorporate green infrastructure provisions into codes, policies, and standard practices through a few essential steps. First, the stormwater management plan review would take place early in the development review process to ensure that green infrastructure practices are thoughtfully incorporated into plans. Next, zoning codes and building codes need to result in the same goals and objectives for green infrastructure implementation. For instance, policies such as

harvesting rainwater for irrigation can be an effective green infrastructure strategy when permissible with building codes. To make sure that green infrastructure policies are meeting water quality and other goals, communities will need to monitor and track implementation and maintenance.

EXPECTED BENEFITS

- Reduced stormwater volume and velocity and fewer stormwater overflow events.
- Less polluted stormwater runoff.
- Lower cost for stormwater management facilities.
- Urban heat island mitigation and reduced energy demand.
- Potential recreational and aesthetic amenities.
- Traffic calming.
- More distinctive communities.
- Increased land values.

STEPS TO IMPLEMENTATION

1. Modest Adjustments

- Add stormwater management requirements and water quality elements to comprehensive plans to recognize and allow green infrastructure stormwater management alternatives in zoning and subdivision regulations.
- Complete the EPA Water Quality Scorecard. The tool gives local governments an idea of the range of green infrastructure policies and which might be right for a specific community.
- Offer zoning upgrades, expedited permitting, reduced stormwater requirements, and other incentives for development proposals that include green infrastructure practices.

US Environmental Protection Agency

This mall, Pompano Fashion Square in Pompano Beach, Florida, is a good example of a parking lot that could be repurposed for green infrastructure.

- Encourage site-planning meetings early in the approval process to review the green infrastructure components of development proposals along with other site planning topics.

- Develop incentives for homeowners to install rain barrels, rain gardens, green roofs, and other green infrastructure.

2. Major Modifications

- Develop a performance standard that requires a system of stormwater management where stormwater infiltrates in ground, is either reused on site and/or evapotranspires, and avoids single-use facilities. Require developers to meet stormwater requirements using green infrastructure practices where appropriate.

- Update the community's stormwater design manual with locally appropriate examples and guidelines for designing, installing, and maintaining green infrastructure.

- Review and change, where necessary, building and zoning codes or other local regulations to ensure that green infra-

structure is legal (e.g., remove restrictions on downspout disconnection and stormwater reuse).

- Take into account rainwater harvesting and reuse when setting the stormwater management requirements for a development.

- Develop or revise stormwater utility bills to include a fee based on impervious services to address combined sewer overflows and offer a fee discount based on the use of green infrastructure techniques.

- Conduct inspections of sites and develop mechanisms to enforce stormwater management plans and maintenance agreements.

3. Wholesale Changes

- Give fiscal credit to developers toward stormwater management requirements for preservation of trees and open space, which help to decrease impervious surfaces and allow for stormwater infiltration.

- Amend stormwater management regulations and development codes to allow off-site stormwater management, especially for infill and redevelopment areas.

- Require green infrastructure bonds or other revenue generation in zoning or subdivision ordinances to ensure proper operation and maintenance of green infrastructure stormwater management facilities.

PRACTICE POINTERS

- Engage local governments in regional stormwater management strategies and coordinate future land use and development decisions for large-scale water quality benefits.

- Ensure that all local government departments/agencies coordinate with one another so that green infrastructure meets multiple community objectives (e.g., allow rain gardens to meet landscaping requirements).

- Enact riparian buffer regulations to protect water resources from nonpoint source pollution, stabilize banks, and provide aquatic and wildlife habitat.

- Consider separate stormwater management requirements for densely developed activity centers and infill sites as opposed to greenfield development. Recognize that impervious cover limits, open space requirements, and on-site detention requirements may be appropriate for large greenfield developments but not for more urban sites. Provide flexibility to allow off-site and regional stormwater management facilities, and give credit for alternative approaches like pervious pavement and green roofs.

- Work with key staff from local agencies such as transportation, planning, and public works to integrate green infrastructure into all codes and ordinances.

EXAMPLES AND REFERENCES

- U.S. EPA. *Water Quality Scorecard.* August 2009. http://www.epa.gov/npdes/pubs/gi_municipal_scorecard.pdf.

- U.S. EPA. *Green Infrastructure Municipal Handbook.* (series of publications) http://cfpub.epa.gov/npdes/greeninfrastructure/munichandbook.cfm.

- U.S. EPA. *Stormwater Management Handbook: Implementing Green Infrastructure in Northern Kentucky Communities.* May 2009. http://www.epa.gov/smartgrowth/sgia_communities.htm#ky.

- U.S. EPA. *Protecting Water Quality with Smart Growth*

Strategies and Natural Stormwater Management in Sussex County, Delaware. January 2009. http://www.epa.gov/smartgrowth/noaa_epa_techasst.htm#6.

- U.S. EPA. "Source Water Protection." http://www.epa.gov/nps/ordinance/sourcewater.htm. Accessed July 22, 2009.

- U.S. EPA. "Stormwater Pollution Prevention Plans for Construction Activities." http://cfpub.epa.gov/npdes/stormwater/swppp.cfm. Accessed July 22, 2009.

- U.S. EPA. *Protecting Water Resources with Higher-Density Development.* January 2006. EPA 231-R-06-001. pp. 23-29. http://www.epa.gov/smartgrowth/water_density.htm.

- Center for Neighborhood Technology. "Green Values Stormwater Toolbox." http://greenvalues.cnt.org. Accessed June 20, 2009.

- City of Portland, Oregon. "General Requirements and Policies." *Stormwater Management Manual.* http://www.portlandonline.com/bes/index.cfm?c=35122&a=55769. Accessed June 22, 2009.

- Santa Clara Valley (California) Urban Runoff Pollution Prevention Program. *Operations and Maintenance of Treatment BMPs.* http://www.scvurppp-w2k.com/om_work-product_links.htm. Accessed June 20, 2009.

- U.S. EPA. "Environmental Management Systems." http://www.epa.gov/ems. Accessed June 22, 2009.

- U.S. EPA. *Reducing Stormwater Costs through Low Impact Development (LID) Strategies and Practices.* December 2007. EPA 841-F-07-006. http://www.epa.gov/owow/nps/lid/costs07.

- City of New York. "Water." PlaNYC. http://www.nyc.gov/html/planyc2030/html/plan/water.shtml. Accessed May 19, 2009.

ADOPT SMART ANNEXATION POLICIES

INTRODUCTION

Communities often have the most influence over development on their edges when land is annexed into a municipality. It is then that the greatest opportunity exists to determine how this land will help the community advance its overall planning goals and to ensure that the public costs of providing infrastructure and services for the annexed area are balanced with potential tax and other revenues from the annexed lands (including any exactions or other requirements).

In most states, municipalities face enormous pressure to annex lands. One of the most important forces driving annexation is communities' desire to increase their tax base, thereby increasing revenues into municipal coffers. Further, in growth areas in many states, municipalities fear that if they do not annex aggressively, their neighbors may, hemming them in and limiting their ability to grow. Finally, in many growth areas, municipalities may believe the only way to ensure that growth in the surrounding region occurs responsibly and according to a plan is to annex areas to gain control over planning, development, and design decision-making before development occurs.

Ad hoc annexation is a major cause and enabler of exurban development and sprawl. Ironically, in many cases, the tax burden from annexed areas may exceed the increase in tax revenues, especially over the long term.

RESPONSE TO THE PROBLEM

The principal policies that successful communities use to handle annexations include:

- Revising local codes to anticipate annexations in the comprehensive planning process and to ensure that annexations are consistent with adopted comprehensive plans;

- Developing intergovernmental processes and agreements—between counties and municipalities, and between neighboring municipalities—to guide and govern planning for physical expansion and annexation; and

- Establishing criteria for the review process leading up to potential annexations, including criteria for fiscal impact analyses.

Because many of the forces driving ad hoc annexation stem from local competition for tax base, communities and regions may also need to work together to rationalize their local taxation systems, including consideration of revenue sharing among jurisdictions.

EXPECTED BENEFITS

- Well-planned, contiguous municipal expansion that benefits the community, supports community character and quality of life, and promotes compact development.

- Creation of communities that are "tax positive"—places that have a logical and fiscally sound annexation of land where services and infrastructure are adequate.

- Focus on intergovernmental collaboration instead of competition for territorial expansion leading to over-extension of municipal boundaries and the resulting scattered, leapfrog development.

- Creation of logical, well-planned communities, instead of ad hoc formation of small incorporated municipalities intended primarily to prevent tax increases associated with annexation.

- Orderly, planned community expansion that accommodates population growth and provides the tax base required to meet the community's objectives.

United States Environmental Protection Agency

This urban growth boundary shows a stark contrast between the developed and undeveloped areas of this community.

STEPS TO IMPLEMENTATION

1. Modest Adjustments

- Establish a code requirement that future annexations be consistent with the community comprehensive plan (or local equivalent), along with a requirement that the comprehensive plan map and describe future potential areas of annexation. These could be developed using a sphere of influence/urban transition area approach, like that used in California's Local Agency Formation Commission, or tiered planning areas like those used by the city of Boulder and Boulder County, Colorado.

- Require future potential annexation areas mapped in the comprehensive plan to include a preliminary identification of anticipated zoning, as well as a preliminary description of how municipal services and infrastructure (e.g., water, sanitary sewer, stormwater, transportation, police, and fire) would be funded in annexed areas. This should be based on community service standards and an assessment of existing conditions and capacities in the mapped areas.

- Require the mapping of potential future annexation areas in the comprehensive plan to identify and evaluate any prime agricultural lands, important wildlife habitat, areas of special ecological value or concern, and any lands contaminated by past industrial or agricultural activities or hazardous materials spills.

- Establish a code requirement that the transportation element of the community comprehensive plan (or local equivalent) identify a future collector and arterial street network for any potential annexation areas mapped in the plan. Require extensions of the existing municipal street network to be mapped to meet minimum internal connectivity standards in any annexed areas, as well as minimum external connectivity with existing and future neighborhoods.

2. Major Modifications

- Adopt fiscal impact analysis requirements for proposed annexations, including criteria for the forecast ratio of revenues to costs. Include provisions for additional fees to rectify imbalances.

- Establish a minimum contiguity requirement for any proposed annexation area. For example, at least 25 percent of the circumference of any proposed annexation must be coterminous with the existing incorporated area, subject to exceptions for bodies of water. An adjunct provision or variation would be to specifically prohibit "flagpole" annexations.[9]

- Develop and adopt joint infrastructure standards (e.g., water, sanitary sewer, stormwater, streets) for a municipality and its surrounding county, or by multiple municipalities and/or counties, to be applied to proposed development in areas that may eventually be annexed into a municipality. This ensures that any development in future annexation areas that occurs prior to annexation is compatible with the annexing community. It also ensures that facilities are designed consistently with standards of the municipalities. This coordination discourages landowners or developers from "shopping" one government against another to obtain the combination of services and fees—which could turn out to be a bad deal for the municipality.

3. Wholesale Changes
(Note: some measures below are in support of code changes, but are not in themselves addressed through the zoning or land development code.)

- Develop an intergovernmental agreement between one or more municipalities and one or more counties providing for development and adoption of a multi-jurisdiction comprehensive plan. Include provisions for identifying areas of potential annexation and provisions for zoning, infrastructure, lands of special concern, and street extensions, similar to the four measures described under Modest Adjustments.

- Develop an intergovernmental agreement between one or more municipalities and one or more counties to guide the annexation process in specific areas, which would be mapped in the agreement. Include provisions addressing infrastructure standards, funding for extension of infrastructure and services, and the approval processes of the affected jurisdictions.

- Develop a regional compact or intergovernmental agreement for revenue sharing to reduce or eliminate the pressure to annex land for municipal budget growth.

The Urban Development Boundary in Miami-Dade County, Florida, illustrates the division between land intended for development and area meant to be preserved.

PRACTICE POINTERS

- Annexation law and policy are among the most controversial aspects of growth management. Many states are changing the laws governing the authority of municipalities to annex land, establishing or revising criteria for annexations, requiring additional review and approval by adjacent counties and municipalities, and providing for oversight by third parties or agencies. The first step for any municipality is to make sure that its ordinances are consistent with state law.

- Issues related to estimating costs of extending infrastructure and municipal services into potential annexation areas are difficult to resolve if there are no agreed-upon standards for the timing, placement, and design of facili-

---

9   Flagpole annexations are connected to a municipality through a narrow strip of land.

ties and services. An important step in addressing annexation policy issues is to work—ideally in cooperation with other area governments—on design and service standards to estimate the cost of providing facilities and services.

- One of the potential benefits of good annexation policy, especially with multiple jurisdictions involved, is avoiding the leapfrogging of suburban subdivisions and commercial projects outside municipal areas.

EXAMPLES AND REFERENCES

- California Association of Local Agency Formation Commission. http://www.calafco.org.

- Local Agency Formation Commission of Monterey County, California. "Sphere of Influence Policies and Criteria." October 2006. http://ooosweb.co.monterey.ca.us/lafco/policy.htm.

- Denver Regional Council of Governments. "Mile High Compact." August 2000. http://www.drcog.org/index.cfm?page=MileHighCompact. Accessed May 13, 2009.

- City of Austin, Texas. Smart Growth Initiative. http://www.ci.austin.tx.us/smartgrowth. Accessed May 31, 2009.

- City of Austin, Texas. *Smart Growth Criteria Matrix*. February 2001. http://www.epa.gov/smartgrowth/scorecards/austin_matrix.pdf.

- Boulder County, Colorado. "Intergovernmental Agreements." http://www.bouldercounty.org/lu/igas/index.htm. Accessed June 12, 2009.

- Larimer County, Colorado. Rural Land Use Center. http://www.co.larimer.co.us/rluc. Accessed June 20, 2009.

- Larimer County, Colorado. Larimer County Urban Area Street Standards. April 2007. http://www.co.larimer.co.us/engineering/GMARdStds/GMARdStds.htm.

- Hinze, S. and Baker, K. *Minnesota's Fiscal Disparities Programs*. Minnesota House of Representatives Research Department. January 2005. http://www.house.leg.state.mn.us/hrd/pubs/fiscaldis.pdf.

ENCOURAGE APPROPRIATE DEVELOPMENT DENSITIES ON THE EDGE

INTRODUCTION

On the periphery of urban areas, suburbs, and small towns, communities' development patterns are often not dense enough to support mixed land uses or transit or to create other efficiencies associated with denser development patterns, such as cost-efficient infrastructure. At the same time, these areas are often too dense for rural areas to maintain a truly rural character. Rural development patterns typically:

- Are supported by limited infrastructure (relying, for instance, on gravel roads and septic systems);

- Cost less to support because they use fewer government services; and

- Preserve large tracts of open space and agricultural lands.

This issue is most relevant to exurban development—areas outside the jurisdictional boundaries of cities and towns. The density is approximately 2 to 4 housing units per gross acre at the more suburban end of the spectrum, and one unit per 20 to 40 acres at the rural end. Many suburban, small town, and county zoning codes and subdivision ordinances allow only these densities. Densities can vary based on regional differences. For instance, Western states will have a different threshold than those in the Southeast.

This low-density development pattern has been one of the fastest growing sectors of the housing market, fueled by a variety of factors, including people moving to rural communities for the quality of life, an expanding second-home market for less expensive vacation homes in small towns, and rural communities' desire to grow. Developers have also found such rural areas to be the "path of least resistance." They are generally able to quickly obtain approvals through a county or rural town's less complicated entitlement procedure.

Land use laws, particularly in the Western states, give extensive rights to large landowners, ranchers, and farmers to develop their properties in the future, typically at lower densities. In these places, low-density residential zoning is the de facto zoning that has been overlaid onto many large tracts of land. This means that many areas that are perceived to be rural are, in fact, zoned for residential development that does not fit a rural context.

The desire to remain rural or maintain a small-town character is a common theme in these communities. Lower densities are often encouraged in the belief that they will help preserve an area's rural character. These densities, however, most frequently translate into low-density, cookie-cutter subdivisions, with streets and homes that are more typical of suburban, rather than rural, communities. The most difficult densities are those in the ½-acre to 5-acre range. The difficulties with these densities include:

- Expensive infrastructure to both provide and maintain to serve a minimal number of units;

- Reliance on septic systems, which have a limited capacity over time;

- A land use pattern that is difficult or impossible to intensify later, as it typically includes individual property owners, making land hard to assemble; and

- Farmland that becomes fragmented by these large-lot homes, which means little possibility of carrying on true agriculture or maintaining farm animals in these areas.

These densities are neither rural nor town-like in their character. Once developed, they are difficult to change and become more difficult to maintain over time.

This aerial from suburban Dallas shows how the "Devil's Density" is built out on the edge of the town at residential density that is not efficient with more compact development patterns.

This type of growth also becomes a jurisdictional, city-versus-county issue. Much of this development pattern is occurring within county jurisdictions at or near city limits because large agricultural properties are being developed under county development procedures. The counties often have minimal regulations and limited resources to plan for, review, or process these types of developments. This has made it difficult to control the implementation of policies and restrictions as well as standards for these developments. Developers often are better equipped than county planning and engineering staff to deal with the various complex issues that arise from these developments.

RESPONSE TO THE PROBLEM

Density that cannot support necessary services is not sustainable on any level—fiscally, environmentally, socially, and for public health. In most places, zoning at one unit per 2 gross acres typically cannot support necessary services. When zoning at this density, communities usually are focused more on the perceived market demand and/or potential tax revenue than on what it will take in infrastructure and other resources to support such a pattern. When communities look at the potential impacts and decipher where they can make improvements through increased densities as well as other zoning changes, they can make their neighborhoods fiscally sound and environmentally sustainable.

Finding a solution takes a balance of strategies, combining those that eliminate the types of densities so persistent where urban and rural communities meet with those that direct unsustainable development patterns away from these areas.

When communities grow, their comprehensive plans should cover only areas that form a natural edge to the community and that will not be expanded beyond or leapfrogged in the future. An example may be an area bordering a creek or other natural open space, which provides a natural barrier to expansion and clearly defines an edge to the community. Another strategy is to continue the town's street pattern to use the infrastructure to its fullest capacity and then end in an agricultural zone at the community's edge. This will better integrate large lots into the community by using them to transition to agricultural uses at the town's periphery.

These remedies only address the properties at a community's edge. The most problematic developments are those that employ unsustainable densities outside these areas as ranches, orchards, and farms are developed. These sites are typically in counties' jurisdictions. Counties and towns, therefore, need to coordinate their planning efforts to minimize the ad hoc development of rural areas and integrate their comprehensive plans to include expansion areas and areas that will be maintained for agriculture or open space. Towns and counties will

This New Jersey farmland is punctuated by a low density residential development creating a conflict between providing services to these homes and preserving agricultural uses.

need to tackle this issue together in a comprehensive manner to address planning, engineering, property ownership, and development issues.

EXPECTED BENEFITS

- Lower infrastructure costs for local and state governments and service providers.

- Preservation of large, contiguous blocks of open space and agricultural lands. This is most critical for protecting habitat corridors and maintaining viable agricultural activities and related businesses.

- Support for downtowns and traditional neighborhood developments, with greater connectivity with the immediate town or city.

- Consistent and connected patterns of development instead of leapfrog growth, which disregards planned boundaries.

- Minimizing the areas that are hamstrung by limited redevelopment potential due to ownership patterns and the lack of opportunities for land assembly.

STEPS TO IMPLEMENTATION

(Note: Several implementation steps from Essential Fix No. 8 that support preferred growth areas also apply to this fix, including agricultural interim holding zones, area-specific impact fees, adequate public facilities ordinances, annexation policies, and urban services areas and boundaries.)

1. Modest Adjustments

- Adopt comprehensive plans that encourage sustainable development patterns in peripheral and exurban areas by redesignating density allocations.

- Amend zoning ordinances to repeal zone districts that allow unsustainable densities at the community's edge.

- Develop design regulations that require connectivity and integration with adjacent neighborhoods and create transitions to adjacent agricultural or undeveloped areas.

2. Major Modifications

- Establish benchmarks for intended densities in comprehensive plans in rural areas (e.g., one unit per 80 acres in some Western states).

- Require minimum densities in areas targeted for growth.

- Require cluster/conservation subdivisions at the community's edge to transition to rural areas. These subdivisions are for edge conditions only, with denser zoning on one side and rural areas on the other.

- Require comprehensive fiscal impact and mitigation analysis for proposed rural developments. Require mitigation measures so that rural developments pay their own way.

- Use the SmartCode to categorize and implement the zoning regulations by classifying an appropriate transect for these urban-rural interface areas and adapting the regulations for the community.

3. Wholesale Changes

- Preserve agricultural viability by zoning for large agriculture-only districts.

- Require mandatory annexation as a condition of development approvals in town impact areas (consider a "no objection" clause that is approved by the property owner when annexation is feasible and desired by the town. This clause will make the annexation process predictable and fair).

- Encourage joint town and county policies that set criteria such as location or size controls to coordinate the development of land instead of insular land use resulting from PUDs. (See Essential Fix No. 3.)

PRACTICE POINTERS

- Depending on the state, land patterns, and types of agriculture, the appropriate acreage for agriculturally zoned parcels will vary.

- Consider how rules related to lot splits or family subdivision rights chart the course for inappropriate densities. Family subdivisions are often used to get around minimum lot size regulations.

- In the past, communities have zoned for economic development and property ownership interests, relying on unsustainable development patterns. Often, smaller towns see fees associated with low-density development, along with construction jobs and retail sales, as economic development. Unfortunately, the cost of maintaining the public infrastructure frequently exceeds the value brought with the short-term economic development.

- Do not allow cluster/conservation subdivisions in areas where true rural development patterns are preferred. Clustered subdivisions disrupt agricultural operations.

- In certain circumstances, land trusts have purchased conservation easements from farmers and ranchers that prohibit development. Selling the easement gives landowners some financial benefit without having to develop their land. This strategy allows landowners to maintain their farms.

- Transfer of Development Rights (TDR) programs may be considered; however, these programs are complex and will be feasible only in specific situations.

EXAMPLES AND REFERENCES

- Duerksen, C. and Snyder, C. *Nature-Friendly Communities. Island Press.* May 2005. pp. 40-50.

- Burchell, R. et al. *Costs of Sprawl—2000.* Transit Cooperative Research Program Report 74. Transportation Research Board. June 2002. pp. 26-31.

- Freedgood, J. et al. *Cost of Community Services Studies: Making the Case for Conservation.* American Farmland Trust. August 2002. pp. 55-60. http://www.farmlandinfo. org/farmland_search/index.cfm?articleID=28415&function=article_view.

- Livingston, A. et al. *The Costs of Sprawl: Fiscal, Environmental, and Quality of Life Impacts of Low-Density Development in the Denver Region.* Environment Colorado. March 2003. pp. 24-29. http://www.environmentcolorado.org/envco-growth.asp?id2=9356.

- Tischler, P. *Analyzing the Fiscal Impacts of Development.* Management Information Service Report No. 20. March 1988. pp. 54-56.

- American Farmland Trust. *Saving American Farmland: What Works.* May 1997. pp. 43-47. http://www.farmland-info.org/farmland_preservation_literature/index. cfm?function=article_view&articleID=29384.

- Bowers, D. "Achieving Sensible Agricultural Zoning to Protect PDR Investment." Presented at "Protecting Farmland at the Fringe." September 2001. http://www. farmlandinfo.org/documents/29520/Achieving_Sensible_ Agricultural_Zoning_full_presentation.pdf.

- County of Marin, California. "Agricultural Element – Executive Summary." *Marin Countywide Plan.* http://www. co.marin.ca.us/depts/cd/main/comdev/advance/cwp/ ag.cfm. Accessed August 11, 2009.

- County of Marin, California. 2007 *Marin Countywide Plan.* 2007. http://www.co.marin.ca.us/depts/CD/main/fm/ TOC.cfm. Accessed August 11, 2009.